Lecture Notes in Physics

The Lecture Notes in Physics

The series Lecture Notes in Physics (LNP), founded in 1969, reports new developments in physics research and teaching – quickly and informally, but with a high quality and the explicit aim to summarize and communicate current knowledge in an accessible way. Books published in this series are conceived as bridging material between advanced graduate textbooks and the forefront of research and to serve three purposes:

- to be a compact and modern up-to-date source of reference on a well-defined topic

- to serve as an accessible introduction to the field to postgraduate students and nonspecialist researchers from related areas

- to be a source of advanced teaching material for specialized seminars, courses and schools

Both monographs and multi-author volumes will be considered for publication. Edited volumes should, however, consist of a very limited number of contributions only. Proceedings will not be considered for LNP.

Volumes published in LNP are disseminated both in print and in electronic formats, the electronic archive being available at springerlink.com. The series content is indexed, abstracted and referenced by many abstracting and information services, bibliographic networks, subscription agencies, library networks, and consortia.

Proposals should be sent to a member of the Editorial Board, or directly to the managing editor at Springer:

Christian Caron
Springer Heidelberg
Physics Editorial Department I
Tiergartenstrasse 17
69121 Heidelberg / Germany
christian.caron@springer.com

C. Bär
K. Fredenhagen (Eds.)

Quantum Field Theory on Curved Spacetimes

Concepts and Mathematical Foundations

Springer

Editors
Christian Bär
Universität Potsdam
Inst. Mathematik
14415 Potsdam
Germany
baer@math.uni-potsdam.de

Klaus Fredenhagen
Universität Hamburg
Inst. Theoretische Physik II
Luruper Chaussee 149
22761 Hamburg
Germany
klaus.fredenhagen@desy.de

Bär C., Fredenhagen K. (Eds.), *Quantum Field Theory on Curved Spacetimes: Concepts and Mathematical Foundations*, Lect. Notes Phys. 786 (Springer, Berlin Heidelberg 2009), DOI 10.1007/978-3-642-02780-2

Lecture Notes in Physics ISSN 0075-8450   e-ISSN 1616-6361
ISBN 978-3-642-26051-3   e-ISBN 978-3-642-02780-2
DOI 10.1007/978-3-642-02780-2
Springer Dordrecht Heidelberg London New York

Cover design: Integra Software Services Pvt. Ltd., Pondicherry

Printed on acid-free paper

Springer is part of Springer Science+Business Media (www.springer.com)

Preface

An outstanding problem of theoretical physics is the incorporation of gravity into quantum physics. After the increasing experimental evidence for the validity of Einstein's theory of general relativity, a theory based on the differential geometry of Lorentzian manifolds, and the discovery of the standard model of elementary particle physics, relying on the formalism of quantum field theory, the question of mutual compatibility of these theoretical concepts gains more and more importance. This becomes in particular urgent in modern cosmology where both theories have to be applied simultaneously.

Early attempts of incorporating gravity into quantum field theory by treating the gravitational field as one of the quantum fields run into conceptual and practical problems. This fact led to rather radical new attempts going beyond the established theories, the most prominent ones being string theory and loop quantum gravity. But after some decades of work a satisfactory theory of quantum gravity is still not available; moreover, there are indications that the original field theoretical approach may be better suited than originally expected.

In particular, due to the weakness of gravitational forces, the back reaction of the spacetime metric to the energy momentum tensor of the quantum fields may be neglected, in a first approximation, and one is left with the problem of quantum field theory on Lorentzian manifolds. Surprisingly, this seemingly modest approach leads to far-reaching conceptual and mathematical problems and to spectacular predictions, the most famous one being the Hawking radiation of black holes.

Quantum field theory on Minkowski space is traditionally based on concepts like vacuum, particles, Fock space, S-matrix, and path integrals. It turns out that these concepts are, in general, not well defined on Lorentzian spacetimes. But commutation relations and field equations remain meaningful. Therefore the algebraic approach to quantum field theory proves to be especially well suited for the formulation of quantum field theory on curved spacetimes.

Ingredients of this approach are the formulation of quantum physics in terms of C^*-algebras, the geometry of Lorentzian manifolds, in particular their causal structure, and linear hyperbolic differential equations where the well posedness of the Cauchy problem plays a distinguished role. These ingredients, however, are sufficient only for the treatment of so-called free fields which satisfy linear field equations. The breakthrough for the treatment of nonlinear theories (on the level of formal power series which is also the state of the art in quantum field theories on Minkowski space) relies on the insight (due to M. Radzikowski) that concepts of microlocal analysis are suited for an incorporation of those features of quantum field theory which are on Minkowski space related to the requirement of positivity of energy.

Another major open problem for long time was to find a replacement for the property of symmetry under the isometry group of Minkowski space which plays a crucial role in traditional quantum field theory. The solution to this problem turned out to require means from category theory. Roughly speaking, symmetry has to be replaced by functoriality, and field theoretical constructions can be considered as natural transformations between appropriate functors. From the point of view of physics, the leading idea is that globally hyperbolic subregions of a spacetime have to be considered as spacetimes in their own right, and the allowed constructions apply to all spacetimes (of the class considered) such that they restrict correctly to sub-spacetimes. This was termed the principle of local covariance. It contains the traditional requirement of covariance under spacetime symmetries and the principle of general covariance of general relativity.

Based on it, the perturbative renormalization of quantum field theory on curved spacetime could be carried through. Perturbative renormalization solves the problem of divergences of naive perturbation theory in interacting quantum field theory. In its standard formulation for Minkowski space it heavily relies on translation symmetry. Its combinatorial, algebraic, and analytic structures have been a source of inspiration for mathematics; in recent times in particular the Connes–Kreimer approach found much interest. For curved spacetime the causal perturbation theory of Epstein and Glaser is better suited. As a result, perturbatively renormalized quantum field theory on curved spacetimes has now the status of a proper generalization of quantum field theory on Minkowski space; and it should be able to describe physics on almost all presently accessible scales. Moreover, compared to the Minkowski space theory which often appears to consist of more or less well-defined cooking recipes, the theory becomes more transparent and its fundamental features become visible.

In October 2007 we organized a compact course on quantum field theory on curved spacetimes at the University of Potsdam. More than 40 participants with varying backgrounds came together to learn about the subject including its mathematical prerequisites. Assuming some basic knowledge of differential geometry and functional analysis on the part of the audience we offered several lecture series introducing C^*-algebras, Lorentzian geometry, the classical theory of linear wave equations, and microlocal analysis. Thus prepared the participants then attended the lecture series on the main topic itself, quantum field theory on curved backgrounds.

This book contains the extended lecture notes of this compact course. The logical dependence is as follows:

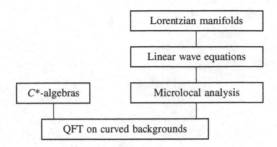

Acknowledgements We are grateful to Sonderforschungsbereich 647 "Raum-Zeit-Materie" and Sonderforschungsbereich 676 "Particles, Strings and the Early Universe" both funded by Deutsche Forschungsgemeinschaft for financially supporting the workshop.

Potsdam, Germany   Christian Bär
Hamburg, Germany   Klaus Fredenhagen

Contents

Chapter 1
C^*-algebras

Christian Bär and Christian Becker

In this chapter we will collect those basic concepts and facts related to C^*-algebras that will be needed later on. We give complete proofs. In Sects. 1.1, 1.2, 1.3, and 1.6 we follow closely the presentation in [1]. For more information on C^*-algebras, see, e.g. [2–6].

1.1 Basic Definitions

Definition 1. *Let A be an associative $\mathbb{C}$-algebra, let $\|\cdot\|$ be a norm on the $\mathbb{C}$-vector space A, and let $* : A \to A$, $a \mapsto a^*$ be a $\mathbb{C}$-antilinear map. Then $(A, \|\cdot\|, *)$ is called a C^*-algebra, if $(A, \|\cdot\|)$ is complete and we have for all $a, b \in A$:*

1. $a^{**} = a$ (*$*$ is an involution*)
2. $(ab)^* = b^* a^*$
3. $\|ab\| \le \|a\| \, \|b\|$ (*submultiplicativity*)
4. $\|a^*\| = \|a\|$ (*$*$ is an isometry*)
5. $\|a^* a\| = \|a\|^2$ (*C^*-property*)

A (not necessarily complete) norm on A satisfying conditions (1) – (5) is called a C^*-norm.

Remark 1. Note that Axioms 1–5 are not independent. For instance, Axiom 4 can easily be deduced from Axioms 1,3, and 5.

Example 1. Let $(H, (\cdot, \cdot))$ be a complex Hilbert space, let $A = \mathcal{L}(H)$ be the algebra of bounded linear operators on H. Let $\|\cdot\|$ be the *operator norm*, i.e.,

$$\|a\| := \sup_{\substack{x \in H \\ \|x\|=1}} \|ax\|.$$

C. Bär (✉)
Institut für Mathematik, Universität Potsdam, Am Neuen Palais 10, D-14469 Potsdam, Germany
e-mail: baer@math.uni-potsdam.de

C. Becker (✉)
Institut für Mathematik, Universität Potsdam, Am Neuen Palais 10, D-14469 Potsdam, Germany
e-mail: becker@math.uni-potsdam.de

Bär, C., Becker, C.: *C*-algebras*. Lect. Notes Phys. **786**, 1–37 (2009)
DOI 10.1007/978-3-642-02780-2_1   © Springer-Verlag Berlin Heidelberg 2009

Let a^* be the operator adjoint to a, i.e.,

$$(ax, y) = (x, a^*y) \qquad \text{for all } x, y \in H.$$

Axioms 1–4 are easily checked. Using Axioms 3 and 4 and the Cauchy–Schwarz inequality we see

$$\|a\|^2 = \sup_{\|x\|=1} \|ax\|^2 = \sup_{\|x\|=1} (ax, ax) = \sup_{\|x\|=1} (x, a^*ax)$$

$$\leq \sup_{\|x\|=1} \|x\| \cdot \|a^*ax\| = \|a^*a\| \overset{\text{Axiom 3}}{\leq} \|a^*\| \cdot \|a\| \overset{\text{Axiom 4}}{=} \|a\|^2.$$

This shows Axiom 5.

Example 2. Let X be a locally compact Hausdorff space. Put

$$A := C_0(X) := \{ f : X \to \mathbb{C} \text{ continuous } \mid \forall \varepsilon > 0 \, \exists K \subset X \text{ compact, so that}$$
$$\forall x \in X \setminus K : |f(x)| < \varepsilon \}.$$

We call $C_0(X)$ the algebra of continuous functions vanishing at infinity. If X is compact, then $A = C_0(X) = C(X)$. All $f \in C_0(X)$ are bounded and we may define

$$\|f\| := \sup_{x \in X} |f(x)|.$$

Moreover let

$$f^*(x) := \overline{f(x)}.$$

Then $(C_0(X), \| \cdot \|, *)$ is a commutative C^*-algebra.

Example 3. Let X be a differentiable manifold. Put

$$A := C_0^\infty(X) := C^\infty(X) \cap C_0(X).$$

We call $C_0^\infty(X)$ the algebra of smooth functions vanishing at infinity. Norm and $*$ are defined as in the previous example. Then $(C_0^\infty(X), \| \cdot \|, *)$ satisfies all axioms of a commutative C^*-algebra except that $(A, \| \cdot \|)$ is not complete. If we complete this normed vector space, then we are back to the previous example of continuous functions.

Definition 2. *A subalgebra A_0 of a C^*-algebra A is called a C^*-subalgebra if it is a closed subspace and $a^* \in A_0$ for all $a \in A_0$.*

Any C^*-subalgebra is a C^*-algebra in its own right.

Definition 3. *Let S be a subset of a C^*-algebra A. Then the intersection of all C^*-subalgebras of A containing S is called the C^*-subalgebra generated by S.*

Definition 4. *An element a of a C^*-algebra is called* self-adjoint *if $a = a^*$*.

Remark 2. Like any algebra a C^*-algebra A has at most one unit 1. Now we have for all $a \in A$

$$1^*a = (1^*a)^{**} = (a^*1^{**})^* = (a^*1)^* = a^{**} = a$$

and similarly one sees $a1^* = a$. Thus 1^* is also a unit. By uniqueness $1 = 1^*$, i.e., the unit is self-adjoint. Moreover,

$$\|1\| = \|1^*1\| = \|1\|^2,$$

hence $\|1\| = 1$ or $\|1\| = 0$. In the second case $1 = 0$ and therefore $A = 0$. Hence we may (and will) from now on assume that $\|1\| = 1$.

Example 4. 1. In Example 1 the algebra $A = \mathcal{L}(H)$ has a unit $1 = \mathrm{id}_H$.
2. The algebra $A = C_0(X)$ has a unit $f \equiv 1$ if and only if $C_0(X) = C(X)$, i.e., if and only if X is compact.

Let A be a C^*-algebra with unit 1. We write $A^\times$ for the set of invertible elements in A. If $a \in A^\times$, then also $a^* \in A^\times$ because

$$a^* \cdot (a^{-1})^* = (a^{-1}a)^* = 1^* = 1,$$

and similarly $(a^{-1})^* \cdot a^* = 1$. Hence $(a^*)^{-1} = (a^{-1})^*$.

Lemma 1. *Let A be a C^*-algebra. Then the maps*

$$\begin{aligned}
A \times A &\to A, &(a, b) &\mapsto a + b, \\
\mathbb{C} \times A &\to A, &(\alpha, a) &\mapsto \alpha a, \\
A \times A &\to A, &(a, b) &\mapsto a \cdot b, \\
A^\times &\to A^\times, &a &\mapsto a^{-1}, \\
A &\to A, &a &\mapsto a^*
\end{aligned}$$

are continuous.

Proof. (a) The first two maps are continuous for all normed vector spaces. This easily follows from the triangle inequality and from homogeneity of the norm.
(b) *Continuity of multiplication.* Let $a_0, b_0 \in A$. Then we have for all $a, b \in A$ with $\|a - a_0\| < \varepsilon$ and $\|b - b_0\| < \varepsilon$:

$$\begin{aligned}
\|ab - a_0b_0\| &= \|ab - a_0b + a_0b - a_0b_0\| \\
&\leq \|a - a_0\| \cdot \|b\| + \|a_0\| \cdot \|b - b_0\| \\
&\leq \varepsilon(\|b - b_0\| + \|b_0\|) + \|a_0\| \cdot \varepsilon \\
&\leq \varepsilon(\varepsilon + \|b_0\|) + \|a_0\| \cdot \varepsilon.
\end{aligned}$$

(c) *Continuity of inversion.* Let $a_0 \in A^\times$. Then we have for all $a \in A^\times$ with $\|a - a_0\| < \varepsilon < \|a_0^{-1}\|^{-1}$

$$\|a^{-1} - a_0^{-1}\| = \|a^{-1}(a_0 - a)a_0^{-1}\|$$
$$\leq \|a^{-1}\| \cdot \|a_0 - a\| \cdot \|a_0^{-1}\|$$
$$\leq \left(\|a^{-1} - a_0^{-1}\| + \|a_0^{-1}\|\right) \cdot \varepsilon \cdot \|a_0^{-1}\|.$$

Thus

$$\underbrace{\left(1 - \varepsilon\|a_0^{-1}\|\right)}_{>0, \text{ since } \varepsilon < \|a_0^{-1}\|^{-1}} \|a^{-1} - a_0^{-1}\| \leq \varepsilon \cdot \|a_0^{-1}\|^2$$

and therefore

$$\|a^{-1} - a_0^{-1}\| \leq \frac{\varepsilon}{1 - \varepsilon\|a_0^{-1}\|} \cdot \|a_0^{-1}\|^2.$$

(d) *Continuity of* $*$ is clear because $*$ is an isometry. $\square$

Remark 3. If $(A, \|\cdot\|, *)$ satisfies the axioms of a C^*-algebra except that $(A, \|\cdot\|)$ is not complete, then the above lemma still holds because completeness has not been used in the proof. Let $\bar{A}$ be the completion of A with respect to the norm $\|\cdot\|$. By the above lemma $+$, $\cdot$, and $*$ extend continuously to $\bar{A}$ thus turning $\bar{A}$ into a C^*-algebra.

1.2 The Spectrum

Definition 5. *Let A be a C^*-algebra with unit 1. For $a \in A$ we call*

$$r_A(a) := \{\lambda \in \mathbb{C} \mid \lambda \cdot 1 - a \in A^\times\}$$

the resolvent set *of a and*

$$\sigma_A(a) := \mathbb{C} \setminus r_A(a)$$

the spectrum *of a. For $\lambda \in r_A(a)$*

$$(\lambda \cdot 1 - a)^{-1} \in A$$

is called the resolvent *of a at λ. Moreover, the number*

$$\rho_A(a) := \sup\{|\lambda| \mid \lambda \in \sigma_A(a)\}$$

is called the spectral radius *of a.*

Example 5. Let X be a compact Hausdorff space and let $A = C(X)$. Then

$$A^\times = \{f \in C(X) \mid f(x) \neq 0 \text{ for all } x \in X\},$$
$$\sigma_{C(X)}(f) = f(X) \subset \mathbb{C},$$
$$r_{C(X)}(f) = \mathbb{C} \setminus f(X),$$
$$\rho_{C(X)}(f) = \|f\|_\infty = \max_{x \in X} |f(x)|.$$

Proposition 1. *Let A be a C^*-algebra with unit 1 and let $a \in A$. Then $\sigma_A(a) \subset \mathbb{C}$ is a nonempty compact subset and the resolvent*

$$r_A(a) \to A, \qquad \lambda \mapsto (\lambda \cdot 1 - a)^{-1}$$

is continuous. Moreover,

$$\rho_A(a) = \lim_{n \to \infty} \|a^n\|^{\frac{1}{n}} = \inf_{n \in \mathbb{N}} \|a^n\|^{\frac{1}{n}} \le \|a\|.$$

Proof. (a) Let $\lambda_0 \in r_A(a)$. For $\lambda \in \mathbb{C}$ with

$$|\lambda - \lambda_0| < \|(\lambda_0 1 - a)^{-1}\|^{-1} \tag{1.1}$$

the Neumann series

$$\sum_{m=0}^{\infty} (\lambda_0 - \lambda)^m (\lambda_0 1 - a)^{-m-1}$$

converges absolutely because

$$\|(\lambda_0 - \lambda)^m (\lambda_0 1 - a)^{-m-1}\| \le |\lambda_0 - \lambda|^m \cdot \|(\lambda_0 1 - a)^{-1}\|^{m+1}$$

$$= \|(\lambda_0 1 - a)^{-1}\| \cdot \underbrace{\left(\frac{\|(\lambda_0 1 - a)^{-1}\|}{|\lambda_0 - \lambda|^{-1}} \right)^m}_{<1 \text{ by } (1.1)}.$$

Since A is complete the Neumann series converges in A. It converges to the resolvent $(\lambda 1 - a)^{-1}$ because

$$(\lambda 1 - a) \sum_{m=0}^{\infty} (\lambda_0 - \lambda)^m (\lambda_0 1 - a)^{-m-1}$$

$$= [(\lambda - \lambda_0)1 + (\lambda_0 1 - a)] \sum_{m=0}^{\infty} (\lambda_0 - \lambda)^m (\lambda_0 1 - a)^{-m-1}$$

$$= - \sum_{m=0}^{\infty} (\lambda_0 - \lambda)^{m+1} (\lambda_0 1 - a)^{-m-1} + \sum_{m=0}^{\infty} (\lambda_0 - \lambda)^m (\lambda_0 1 - a)^{-m}$$

$$= 1.$$

Thus we have shown $\lambda \in r_A(a)$ for all λ satisfying (1.1). Hence $r_A(a)$ is open and $\sigma_A(a)$ is closed.

(b) *Continuity of the resolvent.* We estimate the difference of the resolvent of a at λ_0 and at λ using the Neumann series. If λ satisfies (1.1), then

$$\left\| (\lambda 1 - a)^{-1} - (\lambda_0 1 - a)^{-1} \right\| = \left\| \sum_{m=0}^{\infty} (\lambda_0 - \lambda)^m (\lambda_0 1 - a)^{-m-1} - (\lambda_0 1 - a)^{-1} \right\|$$

$$\leq \sum_{m=1}^{\infty} |\lambda_0 - \lambda|^m \, \| (\lambda_0 1 - a)^{-1} \|^{m+1}$$

$$= \| (\lambda_0 1 - a)^{-1} \| \cdot \frac{|\lambda_0 - \lambda| \cdot \| (\lambda_0 1 - a)^{-1} \|}{1 - |\lambda_0 - \lambda| \cdot \| (\lambda_0 1 - a)^{-1} \|}$$

$$= |\lambda_0 - \lambda| \cdot \frac{\| (\lambda_0 1 - a)^{-1} \|^2}{1 - |\lambda_0 - \lambda| \cdot \| (\lambda_0 1 - a)^{-1} \|}$$

$$\to 0 \quad \text{for } \lambda \to \lambda_0.$$

Hence the resolvent is continuous.

(c) We show $\rho_A(a) \leq \inf_n \|a^n\|^{\frac{1}{n}} \leq \liminf_{n\to\infty} \|a^n\|^{\frac{1}{n}}$. Let $n \in \mathbb{N}$ be fixed and let $|\lambda|^n > \|a^n\|$. Each $m \in \mathbb{N}_0$ can be written uniquely in the form $m = pn + q$, p, $q \in \mathbb{N}_0$, $0 \leq q \leq n - 1$. The series

$$\frac{1}{\lambda} \sum_{m=0}^{\infty} \left(\frac{a}{\lambda} \right)^m = \frac{1}{\lambda} \sum_{q=0}^{n-1} \left(\frac{a}{\lambda} \right)^q \sum_{p=0}^{\infty} \underbrace{\left(\frac{a^n}{\lambda^n} \right)^p}_{\|\cdot\| < 1}$$

converges absolutely. Its limit is $(\lambda 1 - a)^{-1}$ because

$$(\lambda 1 - a) \cdot \left(\sum_{m=0}^{\infty} \lambda^{-m-1} a^m \right) = \sum_{m=0}^{\infty} \lambda^{-m} a^m - \sum_{m=0}^{\infty} \lambda^{-m-1} a^{m+1} = 1$$

and similarly

$$\left(\sum_{m=0}^{\infty} \lambda^{-m-1} a^m \right) \cdot (\lambda 1 - a) = 1.$$

Hence for $|\lambda|^n > \|a^n\|$ the element $(\lambda 1 - a)$ is invertible and thus $\lambda \in r_A(a)$. Therefore

$$\rho_A(a) \leq \inf_{n\in\mathbb{N}} \|a^n\|^{\frac{1}{n}} \leq \liminf_{n\to\infty} \|a^n\|^{\frac{1}{n}}.$$

(d) We show $\rho_A(a) \geq \limsup_{n\to\infty} \|a^n\|^{\frac{1}{n}}$. We abbreviate $\tilde{\rho}(a) := \limsup_{n\to\infty} \|a^n\|^{\frac{1}{n}}$.
Case 1: $\tilde{\rho}(a) = 0$. If a were invertible, then

$$1 = \|1\| = \|a^n a^{-n}\| \leq \|a^n\| \cdot \|a^{-n}\|$$

would imply $1 \leq \tilde{\rho}(a) \cdot \tilde{\rho}(a^{-1}) = 0$, which would yield a contradiction. Therefore $a \notin A^{\times}$. Thus $0 \in \sigma_A(a)$. In particular, the spectrum of a is nonempty. Hence the spectral radius $\rho_A(a)$ is bounded from below by 0 and thus

$$\tilde{\rho}(a) = 0 \le \rho_A(a).$$

Case 2: $\tilde{\rho}(a) > 0$. If $a_n \in A$ are elements for which $R_n := (1 - a_n)^{-1}$ exist, then

$$a_n \to 0 \quad \Leftrightarrow \quad R_n \to 1.$$

This follows from the fact that the map $A^\times \to A^\times$, $a \mapsto a^{-1}$ is continuous by Lemma 1. Put

$$S := \{\lambda \in \mathbb{C} \mid |\lambda| \ge \tilde{\rho}(a)\}.$$

We want to show that $S \not\subset r_A(a)$ since then there exists $\lambda \in \sigma_A(a)$ such that $|\lambda| \ge \tilde{\rho}(a)$ and hence

$$\rho_A(a) \ge |\lambda| \ge \tilde{\rho}(a).$$

Assume in the contrary that $S \subset r_A(a)$. Let $\omega \in \mathbb{C}$ be an nth root of unity, i.e., $\omega^n = 1$. For $\lambda \in S$ we also have $\lambda / \omega^k \in S \subset r_A(a)$. Hence there exists

$$\left(\frac{\lambda}{\omega^k} 1 - a\right)^{-1} = \frac{\omega^k}{\lambda}\left(1 - \frac{\omega^k a}{\lambda}\right)^{-1}$$

and we may define

$$R_n(a, \lambda) := \frac{1}{n} \sum_{k=1}^{n} \left(1 - \frac{\omega^k a}{\lambda}\right)^{-1}.$$

We compute

$$
\begin{aligned}
\left(1 - \frac{a^n}{\lambda^n}\right) R_n(a, \lambda) &= \frac{1}{n} \sum_{k=1}^{n} \sum_{l=1}^{n} \left(\frac{\omega^{k(l-1)} a^{l-1}}{\lambda^{l-1}} - \frac{\omega^{kl} a^l}{\lambda^l}\right)\left(1 - \frac{\omega^k a}{\lambda}\right)^{-1} \\
&= \frac{1}{n} \sum_{k=1}^{n} \sum_{l=1}^{n} \frac{\omega^{k(l-1)} a^{l-1}}{\lambda^{l-1}} \\
&= \frac{1}{n} \sum_{l=1}^{n} \frac{a^{l-1}}{\lambda^{l-1}} \underbrace{\sum_{k=1}^{n} (\omega^{l-1})^k}_{= \begin{cases} 0 & \text{if } l \ge 2 \\ n & \text{if } l = 1 \end{cases}} \\
&= 1.
\end{aligned}
$$

Similarly one sees $R_n(a, \lambda)\left(1 - \frac{a^n}{\lambda^n}\right) = 1$. Hence

$$R_n(a, \lambda) = \left(1 - \frac{a^n}{\lambda^n}\right)^{-1}$$

for any $\lambda \in S \subset r_A(a)$. Moreover for $\lambda \in S$ we have

$$\left\| \left(1 - \frac{a^n}{\tilde{\rho}(a)^n}\right)^{-1} - \left(1 - \frac{a^n}{\lambda^n}\right)^{-1} \right\|$$

$$\leq \frac{1}{n} \sum_{k=1}^{n} \left\| \left(1 - \frac{\omega^k a}{\tilde{\rho}(a)}\right)^{-1} - \left(1 - \frac{\omega^k a}{\lambda}\right)^{-1} \right\|$$

$$= \frac{1}{n} \sum_{k=1}^{n} \left\| \left(1 - \frac{\omega^k a}{\tilde{\rho}(a)}\right)^{-1} \left(1 - \frac{\omega^k a}{\lambda} - 1 + \frac{\omega^k a}{\tilde{\rho}(a)}\right) \left(1 - \frac{\omega^k a}{\lambda}\right)^{-1} \right\|$$

$$= \frac{1}{n} \sum_{k=1}^{n} \left\| \left(\frac{\tilde{\rho}(a)}{\omega^k} 1 - a\right)^{-1} \left(-\frac{\tilde{\rho}(a) a}{\omega^k} + \frac{\lambda a}{\omega^k}\right) \left(\frac{\lambda}{\omega^k} 1 - a\right)^{-1} \right\|$$

$$\leq |\tilde{\rho}(a) - \lambda| \cdot \|a\| \cdot \sup_{z \in S} \|(z1 - a)^{-1}\|^2.$$

The supremum is finite since $z \mapsto (z1 - a)^{-1}$ is continuous on $r_A(a) \supset S$ by part (b) of the proof and since for $|z| \geq 2 \cdot \|a\|$ we have

$$\|(z1 - a)^{-1}\| \leq \frac{1}{|z|} \sum_{n=0}^{\infty} \underbrace{\frac{\|a\|^n}{|z|^n}}_{\leq (\frac{1}{2})^n} \leq \frac{2}{|z|} \leq \frac{1}{\|a\|}.$$

Outside the annulus $\overline{B}_{2\|a\|}(0) - B_{\tilde{\rho}(a)}(0)$ the expression $\|(z1 - a)^{-1}\|$ is bounded by $1 / \|a\|$ and on the compact annulus it is bounded by continuity. Put

$$C := \|a\| \cdot \sup_{z \in S} \|(z1 - a)^{-1}\|^2.$$

We have shown

$$\|R_n(a, \tilde{\rho}(a)) - R_n(a, \lambda)\| \leq C \cdot |\tilde{\rho}(a) - \lambda|$$

for all $n \in \mathbb{N}$ and all $\lambda \in S$. Putting $\lambda = \tilde{\rho}(a) + \frac{1}{j}$ we obtain

$$\left\| \left(1 - \frac{a^n}{\tilde{\rho}(a)^n}\right)^{-1} - \underbrace{\left(1 - \underbrace{\frac{a^n}{(\tilde{\rho}(a) + \frac{1}{j})^n}}_{\to 0 \text{ for } n \to \infty}\right)^{-1}}_{\to 1 \text{ for } n \to \infty} \right\| \leq \frac{C}{j},$$

thus

$$\limsup_{n\to\infty}\left\|\left(1-\frac{a^n}{\widetilde{\rho}(a)^n}\right)^{-1}-1\right\|\le\frac{C}{j}$$

for all $j\in\mathbb{N}$ and hence

$$\limsup_{n\to\infty}\left\|\left(1-\frac{a^n}{\widetilde{\rho}(a)^n}\right)^{-1}-1\right\|=0.$$

For $n\to\infty$ we get

$$\left(1-\frac{a^n}{\widetilde{\rho}(a)^n}\right)^{-1}\to 1$$

and thus

$$\frac{\|a^n\|}{\widetilde{\rho}(a)^n}\to 0. \tag{1.2}$$

On the other hand we have

$$\|a^{n+1}\|^{\frac{1}{n+1}}\le\|a\|^{\frac{1}{n+1}}\cdot\|a^n\|^{\frac{1}{n+1}}$$
$$=\|a\|^{\frac{1}{n+1}}\cdot\|a^n\|^{-\frac{1}{n(n+1)}}\cdot\|a^n\|^{\frac{1}{n}}$$
$$\le\|a\|^{\frac{1}{n+1}}\cdot\|a\|^{-\frac{n}{n(n+1)}}\cdot\|a^n\|^{\frac{1}{n}}$$
$$=\|a^n\|^{\frac{1}{n}}.$$

Hence the sequence $\left(\|a^n\|^{\frac{1}{n}}\right)_{n\in\mathbb{N}}$ is monotonically nonincreasing and therefore

$$\widetilde{\rho}(a)=\limsup_{k\to\infty}\|a^k\|^{\frac{1}{k}}\le\|a^n\|^{\frac{1}{n}}\qquad\text{for all }n\in\mathbb{N}.$$

Thus $1\le\|a^n\|/\widetilde{\rho}(a)^n$ for all $n\in\mathbb{N}$, in contradiction to (1.2).

(e) *The spectrum is nonempty.* If $\sigma(a)=\emptyset$, then $\rho_A(a)=-\infty$ contradicting $\rho_A(a)=\lim_{n\to\infty}\|a^n\|^{\frac{1}{n}}\ge 0$. $\qquad\square$

Definition 6. *Let A be a C^*-algebra with unit. Then $a\in A$ is called*

- normal, *if $aa^*=a^*a$,*
- an isometry, *if $a^*a=1$, and*
- unitary, *if $a^*a=aa^*=1$.*

Remark 4. In particular, self-adjoint elements are normal. In a commutative algebra all elements are normal.

Proposition 2. *Let A be a C^*-algebra with unit and let $a,b\in A$. Then the following holds:*

1. $\sigma_A(a^*) = \overline{\sigma_A(a)} = \{\lambda \in \mathbb{C} \mid \overline{\lambda} \in \sigma_A(a)\}$.
2. If $a \in A^\times$, then $\sigma_A(a^{-1}) = \sigma_A(a)^{-1}$.
3. If a is normal, then $\rho_A(a) = \|a\|$.
4. If a is an isometry, then $\rho_A(a) = 1$.
5. If a is unitary, then $\sigma_A(a) \subset S^1 \subset \mathbb{C}$.
6. If a is self-adjoint, then $\sigma_A(a) \subset [-\|a\|, \|a\|]$ and moreover $\sigma_A(a^2) \subset [0, \|a\|^2]$.
7. If $P(z)$ is a polynomial with complex coefficients and $a \in A$ is arbitrary, then

$$\sigma_A\big(P(a)\big) = P\big(\sigma_A(a)\big) = \{P(\lambda) \mid \lambda \in \sigma_A(a)\}.$$

8. $\sigma_A(ab) - \{0\} = \sigma_A(ba) - \{0\}$.

Proof. We start by showing Assertion 1. A number λ does not lie in the spectrum of a if and only if $(\lambda 1 - a)$ is invertible, i.e., if and only if $(\lambda 1 - a)^* = \overline{\lambda} 1 - a^*$ is invertible, i.e., if and only if $\overline{\lambda}$ does not lie in the spectrum of a^*.

To see Assertion 2 let a be invertible. Then 0 lies neither in the spectrum $\sigma_A(a)$ of a nor in the spectrum $\sigma_A(a^{-1})$ of a^{-1}. Moreover, we have for $\lambda \neq 0$

$$\lambda 1 - a = \lambda a(a^{-1} - \lambda^{-1} 1)$$

and

$$\lambda^{-1} 1 - a^{-1} = \lambda^{-1} a^{-1}(a - \lambda 1).$$

Hence $\lambda 1 - a$ is invertible if and only if $\lambda^{-1} 1 - a^{-1}$ is invertible.

To show Assertion 3 let a be normal. Then a^*a is self-adjoint, in particular normal. Using the C^*-property we obtain inductively

$$\begin{aligned}
\|a^{2^n}\|^2 &= \|(a^{2^n})^* a^{2^n}\| = \|(a^*)^{2^n} a^{2^n}\| = \|(a^*a)^{2^n}\| \\
&= \|(a^*a)^{2^{n-1}} (a^*a)^{2^{n-1}}\| = \|(a^*a)^{2^{n-1}}\|^2 \\
&= \cdots = \|a^*a\|^{2^n} = \|a\|^{2^{n+1}}.
\end{aligned}$$

Thus

$$\rho_A(a) = \lim_{n \to \infty} \|a^{2^n}\|^{\frac{1}{2^n}} = \lim_{n \to \infty} \|a\| = \|a\|.$$

To prove Assertion 4 let a be an isometry. Then

$$\|a^n\|^2 = \|(a^n)^* a^n\| = \|(a^*)^n a^n\| = \|1\| = 1.$$

Hence

$$\rho_A(a) = \lim_{n \to \infty} \|a^n\|^{\frac{1}{n}} = 1.$$

For Assertion 5 let a be unitary. On the one hand we have by Assertion 4

$$\sigma_A(a) \subset \{\lambda \in \mathbb{C} \mid |\lambda| \leq 1\}.$$

On the other hand we have

$$\sigma_A(a) \overset{(1)}{=} \overline{\sigma_A(a^*)} = \overline{\sigma_A(a^{-1})} \overset{(2)}{=} \overline{\sigma_A(a)}^{-1}.$$

Both combined yield $\sigma_A(a) \subset S^1$.

To show Assertion 6 let a be self-adjoint. We need to show $\sigma_A(a) \subset \mathbb{R}$. Let $\lambda \in \mathbb{R}$ with $\lambda^{-1} > \|a\|$. Then $|-i\lambda^{-1}| = \lambda^{-1} > \rho(a)$ and hence $1 + i\lambda a = i\lambda(-i\lambda^{-1} + a)$ is invertible. Put

$$U := (1 - i\lambda a)(1 + i\lambda a)^{-1}.$$

Then $U^* = ((1 + i\lambda a)^{-1})^*(1 - i\lambda a)^* = (1 - i\lambda a^*)^{-1} \cdot (1 + i\lambda a^*) = (1 - i\lambda a)^{-1} \cdot (1 + i\lambda a)$ and therefore

$$
\begin{aligned}
U^*U &= (1 - i\lambda a)^{-1} \cdot (1 + i\lambda a)(1 - i\lambda a)(1 + i\lambda a)^{-1} \\
&= (1 - i\lambda a)^{-1}(1 - i\lambda a)(1 + i\lambda a)(1 + i\lambda a)^{-1} \\
&= 1.
\end{aligned}
$$

Similarly $UU^* = 1$, i.e., U is unitary. By Assertion 5 $\sigma_A(U) \subset S^1$. A simple computation with complex numbers shows that

$$|(1 - i\lambda\mu)(1 + i\lambda\mu)^{-1}| = 1 \quad \Leftrightarrow \quad \mu \in \mathbb{R}.$$

Thus $(1 - i\lambda\mu)(1 + i\lambda\mu)^{-1} \cdot 1 - U$ is invertible if $\mu \in \mathbb{C} \setminus \mathbb{R}$. From

$$
\begin{aligned}
&(1 - i\lambda\mu)(1 + i\lambda\mu)^{-1} \cdot 1 - U \\
&= (1 + i\lambda\mu)^{-1}\big((1 - i\lambda\mu)(1 + i\lambda a)1 - (1 + i\lambda\mu)(1 - i\lambda a)\big)(1 + i\lambda a)^{-1} \\
&= 2i\lambda(1 + i\lambda\mu)^{-1}(a - \mu 1)(1 + i\lambda a)^{-1}
\end{aligned}
$$

we see that $a - \mu 1$ is invertible for all $\mu \in \mathbb{C} \setminus \mathbb{R}$. Thus $\mu \in r_A(a)$ for all $\mu \in \mathbb{C} \setminus \mathbb{R}$ and hence $\sigma_A(a) \subset \mathbb{R}$. The statement about $\sigma_A(a^2)$ now follows from part 7.

To prove Assertion 7 decompose the polynomial $P(z) - \lambda$ into linear factors

$$P(z) - \lambda = \alpha \cdot \prod_{j=1}^{n}(\alpha_j - z), \qquad \alpha, \alpha_j \in \mathbb{C}.$$

We insert an algebra element $a \in A$:

$$P(a) - \lambda 1 = \alpha \cdot \prod_{j=1}^{n}(\alpha_j 1 - a).$$

Since the factors in this product commute the product is invertible if and only if all factors are invertible.[1] In our case this means

$$\lambda \in \sigma_A\big(P(a)\big) \Leftrightarrow \text{at least one factor is noninvertible}$$
$$\Leftrightarrow \alpha_j \in \sigma_A(a) \text{ for some } j$$
$$\Leftrightarrow \lambda = P(\alpha_j) \in P\big(\sigma_A(a)\big).$$

If c is inverse to $1 - ab$, then $(1 + bca) \cdot (1 - ba) = 1 - ba + bc(1 - ab)a = 1$ and $(1 - ba) \cdot (1 + bca) = 1 - ba + b(1 - ab)ca = 1$. Hence $1 + bca$ is inverse to $1 - ba$, which finally yields Assertion 8.   □

Corollary 1. *Let $(A, \| \cdot \|, *)$ be a C*-algebra with unit. Then the norm $\| \cdot \|$ is uniquely determined by A and $*$.*

Proof. For $a \in A$ the element a^*a is self-adjoint and hence

$$\|a\|^2 = \|a^*a\| \overset{2(3)}{=} \rho_A(a^*a)$$

depends only on A and $*$.   □

1.3 Morphisms

Definition 7. *Let A and B be C*-algebras. An algebra homomorphism*

$$\pi : A \to B$$

is called $*$-morphism if for all $a \in A$ we have

$$\pi(a^*) = \pi(a)^*.$$

A map $\pi : A \to A$ is called $*$-automorphism if it is an invertible $*$-morphism.

Corollary 2. *Let A and B be C*-algebras with unit. Each unit-preserving $*$-morphism $\pi : A \to B$ satisfies*

$$\|\pi(a)\| \leq \|a\|$$

for all $a \in A$. In particular, π is continuous.

Proof. For $a \in A^\times$

$$\pi(a)\pi(a^{-1}) = \pi(aa^{-1}) = \pi(1) = 1$$

---

[1] This is generally true in algebras with unit. Let $b = a_1 \cdots a_n$ with commuting factors. Then b is invertible if all factors are invertible: $b^{-1} = a_n^{-1} \cdots a_1^{-1}$. Conversely, if b is invertible, then $a_i^{-1} = b^{-1} \cdot \prod_{j \neq i} a_j$ where we have used that the factors commute.

holds and similarly $\pi(a^{-1})\pi(a) = 1$. Hence $\pi(a) \in B^\times$ with $\pi(a)^{-1} = \pi(a^{-1})$. Now if $\lambda \in r_A(a)$, then

$$\lambda 1 - \pi(a) = \pi(\lambda 1 - a) \in \pi(A^\times) \subset B^\times,$$

i.e., $\lambda \in r_B(\pi(a))$. Hence $r_A(a) \subset r_B(\pi(a))$ and $\sigma_B(\pi(a)) \subset \sigma_A(a)$. This implies the inequality

$$\rho_B(\pi(a)) \leq \rho_A(a).$$

Since π is a $*$-morphism and a^*a and $\pi(a)^*\pi(a)$ are self-adjoint we can estimate the norm as follows:

$$\|\pi(a)\|^2 = \|\pi(a)^*\pi(a)\| = \rho_B(\pi(a)^*\pi(a)) = \rho_B(\pi(a^*a))$$
$$\leq \rho_A(a^*a) = \|a\|^2.$$

$\square$

Corollary 3. *Let A be a C^*-algebra with unit. Then each unit-preserving $*$-automorphism $\pi : A \to A$ satisfies for all $a \in A$:*

$$\|\pi(a)\| = \|a\|.$$

Proof.

$$\|\pi(a)\| \leq \|a\| = \|\pi^{-1}(\pi(a))\| \leq \|\pi(a)\|.$$

$\square$

If $P(z) = \sum_{j=0}^n c_j z^j$ is a polynomial of one complex variable and a an element of an algebra A, then $P(a) = \sum_{j=0}^n c_j a^j$ is defined in an obvious manner. We now show how to define $f(a)$ if f is a continuous function and a is a normal element of a C^*-algebra A. This is known as *continuous functional calculus*.

Proposition 3. *Let A be a C^*-algebra with unit. Let $a \in A$ be normal.*
 Then there is a unique $*$-morphism $C(\sigma_A(a)) \to A$ denoted by $f \mapsto f(a)$ such that $f(a)$ has the standard meaning in case f is the restriction of a polynomial. Moreover, the following holds:

1. $\|f(a)\| = \|f\|_{C(\sigma_A(a))}$ *for all $f \in C(\sigma_A(a))$.*
2. *If B is another C^*-algebra with unit and $\pi : A \to B$ a unit-preserving $*$-morphism, then $\pi(f(a)) = f(\pi(a))$ for all $f \in C(\sigma_A(a))$.*
3. $\sigma_A(f(a)) = f(\sigma_A(a))$ *for all $f \in C(\sigma_A(a))$.[2]*

---

[2] Recall from the proof of Corollary 2 that $\sigma_B(\pi(a)) \subset \sigma_A(a)$. Strictly speaking, the statement is $\pi(f(a)) = (f|_{\sigma_B(\pi(a))})(\pi(a))$.

Proof. For any polynomial P we have that $P(a)$ is also normal and hence by Proposition 2

$$\|P(a)\| = \rho_A(P(a)) = \sup\{|\mu| \mid \mu \in \sigma_A(P(a))\}$$
$$= \sup\{|P(\lambda)| \mid \lambda \in \sigma_A(a)\} = \|P\|_{C(\sigma_A(a))}. \tag{1.3}$$

Thus the map $P \mapsto P(a)$ extends uniquely to a linear map from the closure of the polynomials in $C(\sigma_A(a))$ to A. Since the polynomials form an algebra containing the unit, containing complex conjugates, and separating points, this closure is all of $C(\sigma_A(a))$ by the Stone–Weierstrass theorem. By continuity this extension is a $*$-morphism and Assertion 1 follows from (1.3).

Assertion 2 clearly holds if f is a polynomial. It then follows for continuous f because π is continuous by Corollary 2.

As to Assertion 3 let $\lambda \in \sigma_A(a)$. Choose polynomials P_n such that $P_n \to f$ in $C(\sigma_A(a))$. By Proposition 2 we have $P_n(\lambda) \in \sigma_A(P_n(a))$, i.e., $P_n(a) - P_n(\lambda) \cdot 1 \notin A^\times$. Since the complement of $A^\times$ is closed we can pass to the limit and we obtain $f(a) - f(\lambda) \cdot 1 \notin A^\times$. Hence $f(\lambda) \in \sigma_A(f(a))$. This shows $f(\sigma_A(a)) \subset \sigma_A(f(a))$. Conversely, let $\mu \notin f(\sigma_A(a))$. Then $g := (f - \mu)^{-1} \in C(\sigma(a))$. From $g(a)(f(a) - \mu \cdot 1) = (f(a) - \mu \cdot 1)g(a) = 1$ one sees $f(a) - \mu \cdot 1 \in A^\times$, thus $\mu \notin \sigma(f(a))$. $\qquad\square$

We extend Corollary 3 to the case where π is injective but not necessarily onto. This is not a direct consequence of Corollary 3 because it is not a priori clear that the image of a $*$-morphism is closed and hence a C^*-algebra in its own right.

Proposition 4. *Let A and B be C^*-algebras with unit. Each injective unit-preserving $*$-morphism $\pi : A \to B$ satisfies*

$$\|\pi(a)\| = \|a\|$$

for all $a \in A$.

Proof. By Corollary 2 we only have to show $\|\pi(a)\| \geq \|a\|$. Once we know this inequality for self-adjoint elements it follows for all $a \in A$ because

$$\|\pi(a)\|^2 = \|\pi(a)^*\pi(a)\| = \|\pi(a^*a)\| \geq \|a^*a\| = \|a\|^2.$$

Assume there exists a self-adjoint element $a \in A$ such that $\|\pi(a)\| < \|a\|$. By Proposition 2, we have $\sigma_A(a) \subset [-\|a\|, \|a\|]$ and $\rho_A(a) = \|a\|$, hence $\|a\| \in \sigma_A(a)$ or $-\|a\| \in \sigma_A(a)$. Similarly, $\sigma_B(\pi(a)) \subset [-\|\pi(a)\|, \|\pi(a)\|]$.

Choose a continuous function $f : [-\|a\|, \|a\|] \to \mathbb{R}$ such that f vanishes on $[-\|\pi(a)\|, \|\pi(a)\|]$ and $f(-\|a\|) = f(\|a\|) = 1$. From Proposition 3 we conclude $\pi(f(a)) = f(\pi(a)) = 0$ because $f|_{\sigma_B(\pi(a))} = 0$ and $\|f(a)\| = \|f\|_{C(\sigma_A(a))} \geq 1$. Thus $f(a) \neq 0$. This contradicts the injectivity of π. $\qquad\square$

Remark 5. Any element a in a C^*-algebra A can be represented as a linear combination $a = a_1 + ia_2$ of self-adjoint elements by setting $a_1 := \frac{1}{2} \cdot (a + a^*)$ and $a_2 := \frac{1}{2i} \cdot (a - a^*)$.

Lemma 2. *Let $a \in A$ be a self-adjoint element in a unital C^*-algebra A. Then the following three statements are equivalent:*

1. $a = b^2$ *for a self-adjoint element $b \in A$.*
2. $a = c^*c$ *for an arbitrary element $c \in A$.*
3. $\sigma_A(a) \subset [0, \infty)$.

Proof. If $a = b^2$ for a self-adjoint element, we have by Proposition 3

$$\sigma_A(a) = \sigma_A(b^2) = \{\lambda^2 \mid \lambda \in \sigma_A(b)\} \subset [0, \infty),$$

which proves the implication "1 $\Rightarrow$ 3."

If $\sigma_A(a) \subset [0, \infty)$, we can define the element $b := \sqrt{a}$ using the continuous functional calculus from Proposition 3. We then have $b^* = b$ and $b^2 = a$, which proves the implication "3 $\Rightarrow$ 1."

The implication "1 $\Rightarrow$ 2" is trivial.

Let $a = c^*c$ and suppose $\sigma_A(-a) \subset [0, \infty)$. By Assertion 8 from Proposition 2, we have $\sigma_A(-cc^*) = \sigma_A(-c^*c) - \{0\} \subset [0, \infty)$. Writing $c = c_1 + ic_2$ with self-adjoint elements c_1, c_2, we find $c^*c + cc^* = 2c_1^2 + 2c_2^2$, hence $c^*c = 2c_1^2 + 2c_2^2 - cc^*$, which implies $\sigma_A(c^*c) \subset [0, \infty)$. Hence $\sigma_A(c^*c) = \{0\}$, which implies $c^*c = a = 0$.

Now suppose $a = c^*c$ for an arbitrary element $c \in A$. Since $a = c^*c$ is self-adjoint and $\sigma_A(a^2) \subset [0, \infty)$, by the continuous functional calculus from Proposition 3, there exists a unique element $|a| := \sqrt{a^2}$ with

$$\sigma_A(d) = \{\sqrt{\lambda} \mid \lambda \in \sigma_A(a^2)\} \subset [0, \infty).$$

By the same argument, the elements $a_+ := \frac{1}{2} \cdot (|a| + a)$ and $a_- := \frac{1}{2} \cdot (|a| - a)$ are self-adjoint and satisfy $\sigma_A(a_i) \subset [0, \infty)$. We then have $a = a_+ - a_-$. Further, for the element $d := ca_-$, we compute

$$-d^*d = -a_-c^*ca_- = -a_-(a_+ - a_-)a_- = -a_-a_+a_- + (a_-)^3 = (a_-)^3,$$

since $a_+a_- = \frac{1}{4}(|a| + a) \cdot (|a| - a) = \frac{1}{4}(|a|^2 - a^2) = 0$. We thus have $\sigma_A(-d^*d) = \sigma_A((a_-)^3) \subset [0, \infty)$, which yields $d = 0$. Hence $c = 0$ or $a_- = 0$, thus $a = a_+$ and $\sigma_A(a) = \sigma_A(a_+) \subset [0, \infty)$. This proves the implication "1 $\Rightarrow$ 3." $\square$

Definition 8. *A self-adjoint element $a \in A$ is called* positive, *if one and hence all of the properties in Lemma 2 hold.*

Remark 6. By the reasoning of the preceding proof, any self-adjoint element $a \in A$ can be represented as a linear combination $a = a_+ - a_-$ with positive elements $a_+ := \frac{1}{2} \cdot (|a| + a)$ and $a_- := \frac{1}{2} \cdot (|a| - a)$ satisfying $a_+a_- = 0$. Combining this observation with Remark 5, we conclude that any $*$-subalgebra of A is spanned by its positive elements (of norm ≤ 1).

1.4 States and Representations

Let $(A, \| \cdot \|, *)$ be a C^*-algebra and H a Hilbert space.

Definition 9. *A representation of A on H is a $*$-morphism $\pi : A \to \mathcal{L}(H)$. A representation is called* faithful, *if π is injective. A subset $U \subset H$ is called* invariant *under A, if*

$$\pi(A)U := \{\pi(a) \cdot u \mid a \in a, u \in U\} \subset U .$$

A representation is called irreducible, *if the only closed vector subspaces of H invariant under A are $\{0\}$ and H.*

Remark 7. Let $\pi_\lambda : A \to \mathcal{L}(H_\lambda)$, $\lambda \in \Lambda$ be representations of A. Then

$$\pi = \bigoplus_{\lambda \in \Lambda} \pi_\lambda : A \to \mathcal{L}(\bigoplus_{\lambda \in \Lambda} H_\lambda),$$
$$\pi(a)\big((x_\lambda)_{\lambda \in \Lambda}\big) = \big(\pi_\lambda(a) \cdot x_\lambda\big)_{\lambda \in \Lambda},$$

is called the *direct sum representation.*

Definition 10. *Two representations $\pi_1 : A \to \mathcal{L}(H_1)$, $\pi_2 : A \to \mathcal{L}(H_2)$ are called* unitarily equivalent, *if there exists a unitary operator $U : H_1 \to H_2$, such that for every $a \in A$:*

$$U \circ \pi_1(a) = \pi_2(a) \circ U .$$

Definition 11. *A vector $\Omega \in H$ is called* cyclic *for a representation π, if*

$$\{\pi(a) \cdot \Omega \mid a \in A\} \subset H$$

is a dense subset.

Example 6. The commutative C^*-algebra $A = C(X)$ of continuous functions on a compact Hausdorff space has a natural representation on the Hilbert space $H = L^2(X)$ by multiplication. The constant function $\Omega = 1$ is a cyclic vector since the continuous functions are dense in $L^2(X)$.

Lemma 3. *If (H, π) is an irreducible representation, then either π is the zero map or every non-zero vector $\Omega \in H$ is cyclic for π.*

Proof. For every vector $\Omega \in H$, the space $\pi(A)\Omega$ is invariant under A, hence its closure is either $\{0\}$ or H. If Ω is non-zero then either $\pi(A)\Omega = \{0\}$, so that the one-dimensional subspace $\mathbb{C} \cdot \Omega$ is invariant under A, whence $H = \mathbb{C} \cdot \Omega$ and $\pi = 0$, or there exists an element $a \in A$ such that $\pi(a)\Omega \neq 0$, so that $\pi(A) \cdot \Omega$ is dense in H and hence Ω is cyclic. $\qquad\square$

Definition 12. *A* state *on a C^*-algebra A is a linear functional $\tau : A \to \mathbb{C}$ with*

1. $\|\tau\| := \sup\{|\tau(a)| \mid a \in A, \|a\| = 1\} = 1$ (*τ has norm* 1).
2. $\tau(a^*a) \geq 0 \; \forall a \in A$ (*τ is positive*).

The set of all states on A is denoted by $S(A)$.

Example 7. Let X be a compact Hausdorff space, $A = C(X)$. Let μ be a Borel probability measure on X, i.e., a measure on the Borel sigma algebra of X with $\int_X d\mu = 1$. Then

$$\tau_\mu : A \to \mathbb{C}$$

$$f \mapsto \int_X f \, d\mu$$

is a state. For instance, the state $\mu_{\delta_{x_0}}$ corresponding to the Dirac measure at x_0 is the evaluation at x_0:

$$\mu_{\delta_{x_0}}(f) = f(x_0).$$

Example 8. On the C^*-algebra $A = \mathrm{Mat}(n \times n; \mathbb{C})$ of complex matrices, we have the state

$$\tau(A) := \frac{1}{n} \cdot \mathrm{tr}(A).$$

Example 9. On $A = \mathcal{L}(H)$, a vector $\Omega \in H$ with $\|\Omega\| = 1$ yields a so-called *vector state*

$$\tau(A) := \langle A \cdot \Omega, \Omega \rangle.$$

Proposition 5. *Let $\tau : A \to \mathbb{C}$ be a state on a C^*-algebra A with unit. Then we have the following:*

1. $A \times A \to \mathbb{C}, (a, b) \mapsto \tau(b^*a)$ *is a positive semi-definite, Hermitian sesquilinear form.*
2. $|\tau(b^*a)|^2 \leq \tau(a^*a) \cdot \tau(b^*b)$ $\forall a, b \in A$ (Cauchy–Schwarz inequality).
3. $\tau(a^*) = \overline{\tau(a)}$ $\forall a \in A$.
4. $|\tau(a)|^2 \leq \tau(a^*a)$ $\forall a \in A$.
5. $\tau(1) = \|\tau\| = 1$.

Proof. It follows immediately from the definitions that the form $(a, b) \mapsto \tau(b^*a)$ is sesquilinear and positive semi-definite. To show that it is Hermitian, we set $c = a \cdot z + b$ for some $z \in \mathbb{C}$ and compute

$$0 \leq \tau(c^*c)$$
$$= \bar{z} \cdot z \cdot \tau(a^*a) + \bar{z} \cdot \tau(a^*b) + z \cdot \tau(b^*a) + \tau(b^*b). \qquad (1.4)$$

It follows that $\operatorname{Im}\big(\bar{z} \cdot \tau(a^*b) + z \cdot \tau(b^*a)\big) = 0$. Setting $z = 1$, we obtain $\operatorname{Im}\tau(a^*b) = -\operatorname{Im}\tau(b^*a)$, setting $z = i$, we obtain $\operatorname{Re}\tau(a^*b) = \operatorname{Re}\tau(b^*a)$. Thus $\tau(a^*b) = \overline{\tau(b^*a)}$.

Setting $z = -\frac{\tau(a^*b)}{\tau(a^*a)}$, (1.4) implies the Cauchy–Schwarz inequality:

$$0 \le \frac{|\tau(a^*b)|^2}{\tau(a^*a)} - \frac{|\tau(a^*b)|^2}{\tau(a^*a)} - \frac{|\tau(a^*b)|^2}{\tau(a^*a)} + \tau(b^*b).$$

Since A has a unit, we have

$$\tau(a^*) = \tau(a^*1) = \overline{\tau(1^*a)} = \overline{\tau(a)}.$$

To show Assertion 4, we compute

$$|\tau(a)|^2 = |\tau(1^*a)|^2 \le \tau(1^*1) \cdot \tau(a^*a) = \tau(1) \cdot \tau(a^*a)$$
$$\le \|\tau\| \cdot \|1\| \cdot \tau(a^*a) \le \tau(a^*a).$$

Using $\tau(1) = \tau(1^*1) \ge 0$ and $\tau(1) \le 1$, we compute

$$|\tau(a)|^2 \le \tau(1^*1) \cdot \tau(a^*a) \le \tau(1) \cdot \|\tau\| \cdot \|a^*a\| = \tau(1) \cdot \|a\|^2.$$

We thus have

$$1 = \|\tau\|^2 \le \sup_{\substack{a \in A \\ a \ne 0}} \frac{|\tau(a)|^2}{\|a\|^2} \le \tau(1),$$

hence $\tau(1) = 1$.   □

Remark 8. The proof of Assertion 5 shows that $\varphi(1) = \|\varphi\|$ holds for every positive linear functional φ.

Corollary 4. *Let $\tau_1, \ldots, \tau_n$ be states and $\lambda_1, \ldots, \lambda_n \ge 0$ with $\sum_{j=1}^n \lambda_j = 1$. Then the convex combination $\tau = \sum_{j=1}^n \lambda_n \cdot \tau_n$ is also a state.*

Let τ_i, $i \in \mathbb{N}$ be states and define $\tau(a) := \lim_{i \to \infty} \tau_i(a)$ provided the limit exists. Then τ is a state.

Proof. For the convex combinations, we have

$$\tau(a^*a) = \sum_{j=1}^n \underbrace{\lambda_j}_{\ge 0} \cdot \underbrace{\tau_j(a^*a)}_{\ge 0} \ge 0$$

and

$$\|\tau\| = \tau(1) = \sum_{j=1}^n \lambda_j \cdot \tau_j(1) = \sum_{j=1}^n \lambda_j = 1.$$

Similarly, for the pointwise convergence, we have

$$\tau(a^*a) = \lim_{i \to \infty} \tau_i(a^*a) \geq 0$$

and

$$\tau(1) = \lim_{i \to \infty} \tau_i(1) = \lim_{i \to \infty} 1 = 1.$$

□

Example 10. Let $\tau_1, \ldots, \tau_n$ be vector states for vectors $\Omega_1, \ldots, \Omega_n \in H$. Then for the state $\tau = \sum_{j=1}^{n} \lambda_j \tau_j$ with $\lambda_j \geq 0$, $\sum_{j=1}^{n} \lambda_j = 1$, we find

$$\tau(a) = \sum_{j=1}^{n} \lambda_j \cdot \tau_j(a) = \sum_{j=1}^{n} \lambda_j \cdot (a \cdot \Omega_j, \Omega_j) = \text{tr}(\varrho \cdot a).$$

Here $\varrho \in \mathcal{L}(H)$ is an operator of finite-dimensional range with eigenvectors Ω_j and eigenvalues λ_j.

More generally, a positive trace class operator $\varrho \in \mathcal{L}(H)$ defines a state τ on $A = \mathcal{L}(H)$ by $\tau(a) := \text{tr}(\varrho \cdot a)$. States of this form are called *normal*.

Lemma 4. *Let τ be a state on a C^*-algebra A. Then the following holds:*

1. $\tau(a^*a) = 0 \Leftrightarrow \tau(ba) = 0$ *for any $b \in A$.*
2. $\tau(b^*a^*ab) \leq \|a^*a\| \cdot \tau(b^*b)$.

Proof. 1. Suppose $\tau(a^*a) = 0$. Then the Cauchy–Schwarz inequality

$$|\tau(ba)|^2 \leq \underbrace{\tau(a^*a)}_{=0} \cdot \tau(bb^*) = 0$$

implies $\tau(ba) = 0$. The other direction is obvious.

2. If $\tau(b^*b) = 0$, then $\tau(cb) = 0$ for any $c \in A$, especially for $c = b^*a^*a$. We thus assume $\tau(b^*b) > 0$ and set $\varrho(c) := \frac{\tau(b^*cb)}{\tau(b^*b)}$. Then ϱ is a positive linear functional with $\|\varrho\| = \varrho(1) = 1$. Hence ϱ is a state, and from Proposition 5 we have $\varrho(a^*a) \leq \|a^*a\|$.

□

From every state τ on a C^*-algebra A we can construct a representation of A by making the product $(b, a) \mapsto \tau(b^*a)$ nondegenerate. By Assertion 1 in Lemma 4, the null space

$$N_\tau := \{a \in A \mid \tau(a^*a) = 0\}$$

is a closed linear subspace of A. By Assertion 2 in Lemma 4, N_τ is a left ideal in A. Therefore, the pairing

$$A/N_\tau \times A/N_\tau \to \mathbb{C},$$
$$([a], [b]) \mapsto \tau(b^*a)$$

is a well-defined Hermitian scalar product. Let H_τ be the completion of the pre-Hilbert space A/N_τ. Then the map

$$\pi_\tau : A \to \mathcal{L}(A/N_\tau),$$
$$\pi_\tau(a) \cdot [b] := [ab]$$

satisfies

$$\|\pi_\tau(a) \cdot [b]\|^2 = \tau(b^*a^*ab) \leq \|a^*a\| \cdot \tau(b^*b) = \|a\|^2 \cdot \|[b]\|^2,$$

so $\|\pi_\tau(a)\| \leq \|a\|$ and $\|\pi_\tau\| \leq 1$. The map π_τ thus extends to a representation

$$\pi_\tau : A \to \mathcal{L}(H_\tau).$$

The scalar product induced by $([a], [b]) \mapsto \tau(b^*a)$ on H_τ will be denoted by $\langle \cdot, \cdot \rangle_\tau$.

Definition 13. *Let τ be a state on a C^*-algebra A. The representation $(H_\tau, \langle \cdot, \cdot \rangle_\tau, \pi_\tau)$ constructed above is called the Gelfand–Naimark–Segal representation or GNS representation in short.*

Example 11. For $A = C(X)$ with a state τ_μ given by a probability measure μ as $\tau_\mu(f) = \int_X f \, d\mu$, the representation space of the GNS representation is $H_{\tau_\mu} = L^2(X, \mu)$.

Remark 9. Let τ be a state on a C^*-algebra A with unit. Then we have the following:

1. The vector $\Omega_\tau := [1] \in H_\tau$ is cyclic for π_τ, since

$$\pi_\tau(A) \cdot \Omega_\tau = A/N_\tau \subset H_\tau$$

is dense.
2. τ can be represented as a vector state on the GNS representation because

$$\tau(a) = \tau(1^*a1) = \langle [a1], [1] \rangle_\tau = \langle \pi_\tau(a) \cdot \Omega_\tau, \Omega_\tau \rangle_\tau.$$

Definition 14. *Let A be a C^*-algebra A. The direct sum representation*

$$\bigoplus_{\tau \in S(A)} \pi_\tau : A \to \mathcal{L}\Big(\bigoplus_{\tau \in S(A)} H_\tau \Big)$$

is called the universal representation *of A.*

Remark 10. The universal representation is faithful. Hence every C^*-algebra A is isomorphic to a subalgebra of the algebra $\mathcal{L}(H)$ of bounded linear operators on a Hilbert space H.

Definition 15. *A state τ on a C^*-algebra A is called* pure, *if for every positive linear functional $\varrho : A \to \mathbb{C}$ with $\varrho(a^*a) \leq \tau(a^*a)\ \forall a \in A$, there exists $\lambda \in [0, 1]$ with $\varrho = \lambda \cdot \tau$.*

Remark 11. A pure state τ cannot be written as a convex combination of different states $\tau_1 \neq \tau_2$. If $\tau = \lambda \cdot \tau_1 + (1 - \lambda) \cdot \tau_2$ with $\lambda \in [0, 1]$, then $\tau \geq \lambda \cdot \tau_1$ implies $\lambda = 0$ and $\tau = \tau_2$ or $\lambda = 1$ and $\tau = \tau_1$.

Example 12. The trace as a state of the algebra $A = \mathrm{Mat}(n \times n; \mathbb{C})$ (see Example 8) is not pure unless $n = 1$, namely it can be written as $\frac{1}{n}\mathrm{tr} = \sum_{i=1}^{n} \frac{1}{n}\tau_i$ where τ_i is the vector state for the ith standard unit vector of $\mathbb{C}^n$.

Definition 16. *Let $S \subset A$ be a subset of a C^*-algebra A. The space $S' := \{a \in A \mid [a, s] = 0\ \forall a \in A, s \in S\}$ is called the* commutant *of S. Here $[a, s] := as - sa$ is the* commutator *of a and s.*

Remark 12. If $S \subset A$ is a $*$-invariant subset, i.e., $S^* := \{s^* \mid s \in S\} \subset S$, then the commutant S' is also $*$-invariant. S' is closed, since for every $s \in S$, the map $A \to A, a \mapsto [a, s]$ is continuous. Hence S' is a C^*-subalgebra of A.

Theorem 1. *Let (H, π) be a representation of a unital C^*-algebra A. Then the following two statements are equivalent:*

1. *π is irreducible.*
2. *$(\pi(A))' = \mathbb{C} \cdot \mathrm{id}_H$.*

Proof. Suppose π is irreducible and $b \in \mathcal{L}(H)$ commutes with all elements of $\pi(A)$. By Remark 5, we may write $b = b_1 + ib_2$ with self-adjoint elements $b_1, b_2 \in \mathcal{L}(H)$. We need to show that $\sigma_A(b_1)$ and $\sigma_A(b_2)$ each consist of a single point. Suppose, to the contrary, that $\sigma_A(b_1)$ contains two different numbers $\lambda \neq \mu$. Then we choose functions $f, g \in C(\sigma_A(b_1))$ such that $f(\lambda) = g(\mu) = 1$ and $f \cdot g = 0$. By the continuous functional calculus from Proposition 3 in the C^*-algebra $(\pi(A))'$, we have $f(b_1) \cdot g(b_1) = (f \cdot g)(b_1) = 0$ and $f(b_1), g(b_1) \neq 0$. Since $g(b_1)$ commutes with every element of $\pi(A)$ and π is irreducible, $g(b_1) \cdot H$ is an A-invariant, dense subspace of H. The vanishing of $f(b_1)$ on this subspace implies $f(b_1) = 0$, which contradicts the fact that the continuous functional calculus from Proposition 3 is an isometry. Thus $\sigma_A(b_1)$ consists of a single point, hence $C(\sigma_A(b_1))$ is one dimensional. Since the continuous functional calculus $C(\sigma_A(b_1)) \to A$ is an isometric embedding with b_1, id_H in its image, we conclude that $b_1 = \lambda \mathrm{id}_H$ for a $\lambda \in \mathbb{C}$. By the same argument, b_2 and hence b is a multiple of the identity.

   Now suppose $(\pi(A))' = \mathbb{C} \cdot \mathrm{id}_H$. Let $K \subset H$ be a closed subspace invariant under A, and let p be the orthogonal projection from H onto K. The invariance property $\pi(A)K \subset K$ yields that p commutes with every operator in $\pi(A)$. Hence p is of the form $p = \lambda \cdot \mathrm{id}_H, \lambda \in \mathbb{C}$. Since p is a projection, $p^2 = p$; thus $\lambda^2 = \lambda$. Hence $K = \{0\}$ or $K = H$. $\qquad\square$

Theorem 2. *Let τ be a state on a C^*-algebra A. Then the following two statements are equivalent:*

1. *τ is a pure state.*
2. *The GNS representation (H_τ, π_τ) is irreducible, i.e., H_τ has no nontrivial closed A-invariant subspace.*

Example 13. The GNS representation of the algebra $A = \mathrm{Mat}(n \times n; \mathbb{C})$ for a vector state for $\Omega \in \mathbb{C}^n$ is the standard representation of A on $\mathbb{C}^n$, hence irreducible. Therefore, such vector states are pure.

Proof (of Theorem 2). Suppose τ is a pure state and $v \in \mathcal{L}(H)$ is a positive element of norm ≤ 1 that commutes with every element in $\pi_\tau(A)$. Then the function

$$\varrho : A \to \mathbb{C}, a \mapsto \langle \pi_\tau(a) \cdot v\Omega_\tau, \Omega_\tau \rangle$$

is a positive linear functional on A, satisfying $\varrho(a^*a) \leq \tau(a^*a)$ for all $a \in A$. Hence $\varrho = \lambda \cdot \tau$ for a $\lambda \in [0, 1]$. Thus for arbitrary $a, b \in A$, we obtain in the pre-Hilbert space A/N_τ:

$$\begin{aligned}
\langle v \cdot (a + N_\tau), (b + N_\tau) \rangle_\tau &= \langle v \cdot \pi_\tau(a)\Omega_\tau, \pi_\tau(b)\Omega_\tau \rangle_\tau \\
&= \langle v \cdot \pi_\tau(b^*a)\Omega_\tau, \Omega_\tau \rangle_\tau \\
&= \varrho(b^*a) \\
&= \lambda \cdot \tau(b^*a) \\
&= \langle \lambda \mathrm{id}_{N_\tau} \cdot (a + N_\tau), (b + N_\tau) \rangle_\tau .
\end{aligned}$$

This implies $v = \lambda \mathrm{id}_{H_\tau}$, since A/N_τ is dense in H_τ. By Proposition 1, we conclude that π_τ is irreducible.

Now suppose that π_τ is irreducible. Let ϱ be a positive linear functional on A such that $\varrho(a^*a) \leq \tau(a^*a)$ for all $a \in A$. Then the pairing

$$(a + N_\tau, b + N_\tau) \mapsto \varrho(b^*a)$$

is a positive semi-definite, Hermitian sesquilinear form on A/N_τ. Being majorized by $\langle \cdot, \cdot \rangle_\tau$, it extends to an inner product $\langle \cdot, \cdot \rangle_\varrho$ on the Hilbert space H_τ. Hence there exists a bounded positive operator $m \in \mathcal{L}(H)$ such that

$$\langle x, y \rangle_\varrho = \langle x, my \rangle_\tau \qquad \forall x, y \in H_\tau .$$

Now the estimate

$$0 \leq \varrho(a^*a) = \langle \pi_\tau(a)\Omega, m\pi_\tau(a)\Omega \rangle_\tau \leq \tau(a^*a) = \langle \pi_\tau(a)\Omega, \pi_\tau(a)\Omega \rangle_\tau$$

yields $\|m\| \leq 1$. For every $a, b, c \in A$, we have

$$\langle \pi_\tau(a)\Omega, m\pi_\tau(b)\pi_\tau(c)\Omega \rangle_\tau = \varrho(a^*bc)$$
$$= \varrho((b^*a)^*c)$$
$$= \langle \pi_\tau(b)^*\pi_\tau(a)\Omega, m\pi_\tau(c)\Omega \rangle_\tau$$
$$= \langle \pi_\tau(a)\Omega, \pi_\tau(b)m\pi_\tau(c)\Omega \rangle_\tau .$$

Hence m commutes with every $\pi(b)$, $b \in A$.

By Theorem 1, m is a multiple of the identity and thus ϱ is a multiple of the state τ. This shows that τ is pure. $\qquad\square$

Lemma 5. *In a unital C^*-algebra A, every state is a pointwise limit of convex combinations of pure states.*

Proof. By Corollary 4, convex combinations of pointwise limits of states are states. Hence $S(A)$ is a bounded closed convex set in the topology of pointwise convergence. By the Banach–Alaoglu theorem from functional analysis, $S(A)$ is thus a compact subset of the closed unit ball in the dual space of A (in the topology of pointwise convergence). The Krein–Milman theorem then implies that $S(A)$ is the closed convex hull of its extreme points, which by Remark 11 contain all pure states.

It remains to show that all extreme points in $S(A)$ are pure. Let $\tau \in S(A)$ be an extreme point of $S(A)$, ϱ a positive linear functional on A satisfying $\varrho(a^*a) \leq \tau(a^*a)$ for all $a \in A$, and suppose $\tau \neq \varrho \neq 0$. Then setting $t = \|\varrho\| \in (0, 1)$, we find

$$\tau = t \cdot \frac{\varrho}{\|\varrho\|} + (1 - t) \cdot \frac{(\tau - \varrho)}{\|\tau - \varrho\|} s,$$

since $\|\tau - \varrho\| = \tau(1) - \varrho(1) = \|\tau\| - \|\varrho\| = 1 - t$ by Remark 8. Hence $\varrho/\|\varrho\| = (\tau - \varrho)/\|\tau - \varrho\| = \tau$, since by assumption τ is an extreme point of $S(A)$. Thus $\varrho = t\tau$ and hence τ is a pure state. $\qquad\square$

Remark 13. The restriction of a pure state to a subalgebra need not be pure. For example, let $A = \mathrm{Mat}(4 \times 4; \mathbb{C})$. Then the vector state τ for the unit vector $\Omega = 2^{-1/2}(1, 0, 0, 1)$ is pure. Now embed $B = \mathrm{Mat}(2 \times 2; \mathbb{C})$ as a subalgebra into A via $b \mapsto \begin{pmatrix} b & 0 \\ 0 & b \end{pmatrix}$. The restriction of τ to B yields the trace state of B which is not pure.

The converse can also happen. If τ_1 is the vector state of A for $(1, 0, 0, 0)$ and τ_2 for $(0, 0, 1, 0)$, then $\tau = \frac{1}{2}\tau_1 + \frac{1}{2}\tau_2$ is not pure as a state of A, but it restricts to a pure state for B (the vector state for $(1, 0)$).

1.5 Product States

In this section, we consider states on the tensor product of C^*-algebras. The norms making the algebraic tensor product into a C^*-algebra are highly nonunique. However, the norm making the algebraic tensor product of Hilbert spaces into a pre-Hilbert space is unique. So it seems natural to study norms on the algebras by means of norms on representation spaces. We will work here with the finest norm topology making the algebraic tensor product of C^*-algebras into a C^*-algebra. Throughout this section, we assume the C^*-algebras in question to have a unit.

Remark 14. Let $(H, \langle \cdot, \cdot \rangle_H)$ and $(K, \langle \cdot, \cdot \rangle_K)$ be Hilbert spaces. Then there is a unique inner product $\langle \cdot, \cdot \rangle$ on the algebraic tensor product of H and K such that

$$\langle x \otimes y, x' \otimes y' \rangle = \langle x, x' \rangle_H \cdot \langle y, y' \rangle_K \qquad \forall x, x' \in H, y, y' \in K .$$

The completion of the algebraic tensor product with respect to this inner product is called the tensor product of the Hilbert spaces and is denoted by $H \otimes K$. Moreover, for bounded operators $a \in \mathcal{L}(H)$ and $b \in \mathcal{L}(K)$, there exists a unique operator $a \otimes b \in \mathcal{L}(H \otimes K)$ such that

$$(a \otimes b)(x \otimes y) = a(x) \otimes b(y) \qquad \forall x \in H, y \in K .$$

This operator satisfies $\|a \otimes b\| = \|a\|_H \cdot \|b\|_K$.

Given two C^*-algebras $(A, \| \cdot \|_A, *)$ and $(B, \| \cdot \|_B, *)$, we want to construct C^*-norms on the algebraic tensor product $A \otimes B$. The simplest way to do so is by using the universal representations. The $\mathbb{C}$-antilinear map $* : A \otimes B \to A \otimes B$ defined by $(a \otimes b)^* := a^* \otimes b^*$ on homogeneous elements and extended bilinearly to $A \otimes B$ makes the algebraic tensor product into an involutive algebra.

Lemma 6. *Let (H, φ) and (K, ψ) be representations of A and B, respectively. Then there is a unique $*$-homomorphism $\pi : A \otimes B \to \mathcal{L}(H \otimes K)$ such that*

$$\pi(a \otimes b) = \varphi(a) \otimes \psi(b) \qquad \forall a \in A, b \in B .$$

Moreover, if the representations φ and ψ are faithful, then so is π.

Proof. The map $A \times B \to \mathcal{L}(H \otimes K), (a, b) \mapsto \varphi(a) \otimes \psi(b)$ is bilinear and thus yields a unique linear map $\pi : A \otimes B \to \mathcal{L}(H \otimes K)$ as claimed, which is indeed a $*$-morphism. If both φ and ψ are injective and $z \in A \otimes B$ satisfies $\pi(z) = 0$, then by writing $z = \sum_{j=1}^n a_j \otimes b_j$ with linearly independent b_j, we conclude $\varphi(a_j) = 0$ for $j = 0, \ldots, n$. Hence $a_j = 0$ for $j = 1, \ldots, n$ and thus $z = 0$. $\qquad \square$

By this lemma, it is natural to make use of the universal representation from Definition 14 to obtain a C^*-norm on the algebraic tensor product.

Definition 17. *Let $(A, \| \cdot \|_A, *)$ and $(B, \| \cdot \|_B, *)$ be C^*-algebras with the universal representations $\pi^A : A \to \mathcal{L}(H)$ and $\pi^B : B \to \mathcal{L}(K)$. The* injective C^*-norm $\| \cdot \|_\iota$ *on the algebraic tensor product is defined by*

$$\|c\|_\iota := \|\pi(c)\| ,$$

where $\pi : A \otimes B \to \mathcal{L}(H \otimes K)$ is the unique $*$-morphism induced by $\pi^A \times \pi^B$ as in Lemma 6. The completion of the algebraic tensor product with respect to the C^*-norm $\| \cdot \|_\iota$ is called the injective C^*-tensor product *and is denoted by $A \otimes_\iota B$.*

Since the $*$-morphism π constructed via the universal representations is injective, $\| \cdot \|_\iota$ is indeed a C^*-norm on $A \otimes B$. Another natural C^*-norm on $A \otimes B$ is constructed by taking the supremum over all C^*-norms. By Remark 2, a unit-preserving $*$-morphism π from the algebraic tensor product $A \otimes B$ to a C^*-algebra C satisfies $\|\pi(x)\| \leq \|x\|_\gamma$ with respect to any C^*-norm $\| \cdot \|_\gamma$ on $A \otimes B$. This yields the following characterization of the maximal C^*-norm on $A \otimes B$:

Definition 18. *Let* $(A, \| \cdot \|_A, *)$ *and* $(B, \| \cdot \|_B, *)$ *be* C^*-*algebras. The projective* C^*-*norm* $\| \cdot \|_\pi$ *on the algebraic tensor product* $A \otimes B$ *is defined by*

$$\|c\|_\pi := \inf \left\{ \sum_{j=1}^{n} \|a_j\|_A \cdot \|b_j\|_B \;\middle|\; c = \sum_{j=1}^{n} a_j \otimes b_j \right\} .$$

The completion of $A \otimes B$ *with respect to the* C^*-*norm* $\| \cdot \|_\pi$ *is called the* projective C^*-*tensor product and is denoted by* $A \otimes_\pi B$.

Remark 15. The projective C^*-norm $\| \cdot \|_\pi$ satisfies $\|a \otimes b\|_\pi = \|a\|_A \cdot \|b\|_B$ for all $a \in A, b \in B$. Clearly, any other C^*-norm $\| \cdot \|_\gamma$ on $A \otimes B$ satisfies $\|c\|_\gamma \leq \|c\|_\pi$ for all $c \in A \otimes B$, hence the projective C^*-norm is maximal among all C^*-norms on $A \otimes B$. One can show that the injective C^*-norm $\| \cdot \|_\iota$ on the other hand is minimal among all C^*-norms on $A \otimes B$.

The projective C^*-tensor product has the following universal property.

Lemma 7. *Let* A, B, *and* C *be* C^*-*algebras and let* $\varphi : A \to C$ *and* $\psi : B \to C$ *be* $*$-*morphisms such that* $\varphi(a)$ *and* $\psi(b)$ *commute for all* $a \in A, b \in B$. *Then there exists a unique* $*$-*morphism* $\pi : A \otimes_\pi B \to C$ *such that*

$$\pi(a \otimes b) = \varphi(a) \cdot \psi(b) \qquad \forall a \in A, b \in B . \tag{1.5}$$

Proof. The bilinear map $A \times B \to C$, $(a, b) \mapsto \varphi(a) \cdot \psi(b)$ induces a unique linear map $\pi : A \otimes B \to C$ satisfying (1.5). This map is a $*$-morphism. The map $\| \cdot \|_\gamma : A \otimes B \to \mathbb{R}$, $c \mapsto \|\pi(c)\|_\gamma$ is a C^*-norm; hence it satisfies $\|c\|_\gamma \leq \|c\|_\pi$ for all $c \in A \otimes B$. Hence π is continuous with respect to the projective C^*-norm and thus uniquely extends from the dense subset $A \otimes B$ to the projective C^*-tensor product $A \otimes_\pi B$. $\qquad\square$

Now we study states on the projective C^*-tensor product $A \otimes_\pi B$. Taking linear functionals $\mu : A \to \mathbb{C}$ and $\nu : B \to \mathbb{C}$, setting

$$(\mu \otimes \nu)(a \otimes b) := \mu(a) \cdot \nu(b)$$

on homogeneous elements and extending bilinearly, we obtain a linear functional on $A \otimes B$. In the projective C^*-norm, we have $\|\mu \otimes \nu\|_\pi = \|\mu\|_A \cdot \|\nu\|_B$. Furthermore, for the homogeneous elements $a \otimes b$, we have

$$(\mu \otimes \nu)\big((a \otimes b)^*(a \otimes b)\big) = (\mu \otimes \nu)(a^*a \otimes b^*b) = \mu(a^*a) \cdot \nu(b^*b) .$$

Hence the functional $\mu \otimes \nu : A \otimes B \to \mathbb{C}$ is positive, if μ and ν are.

Definition 19. *Let A and B be C^*-algebras and let $\mu \in S(A)$ and $\nu \in S(B)$ be states. The unique extension of $\mu \otimes \nu$ to a state on the projective C^*-tensor product $A \otimes_\pi B$ is called a* product state.

Since A and B have a unit, we can restrict a state $\tau \in S(A \otimes B)$ to one of the factors by setting

$$\tau^A(a) := \tau(a \otimes 1) \qquad \forall a \in A$$
$$\tau^B(b) := \tau(1 \otimes b) \qquad \forall b \in B$$

Obviously, for any two states $\mu \in S(A)$ and $\nu \in S(B)$, there is a state $\tau \in S(A \otimes_\pi B)$ such that $\tau^A = \mu$ and $\tau^B = \nu$, namely the product state $\tau = \mu \otimes \nu$. Hence in this case, $\tau = \tau^A \otimes \tau^B$, i.e., the measurement in the state τ of an observable in $A \otimes_\pi B$ simply results in the product of measurements in the states τ^A and τ^B, respectively. In general this is not the case, so we set the following.

Definition 20. *A state $\tau \in S(A \otimes_\pi B)$ is called* correlated, *if there exists $a \in A$ and $b \in B$ such that $\tau(a \otimes b) \neq \tau^A(a) \cdot \tau^B(b)$.*

Definition 21. *A state $\tau \in S(A \otimes_\pi B)$ is called* decomposable, *if it is the pointwise limit of convex combinations of product states. A state $\tau \in S(A \otimes_\pi B)$ is called* entangled, *if it is not decomposable.*

Remark 16. In the literature, the pointwise limit of linear functionals is referred to as the weak-* limit. Stated this way, the set of decomposable states is the weak-* closure of the convex hull of the product states.

Example 14. A pure state on $A \otimes_\pi B$ cannot be written as convex combination of different states. Nor can it be written as a pointwise limit of such convex combinations. Hence a pure state is decomposable if and only if it is a product state.

The set of decomposable states is a convex subset of the set of all (positive) linear functionals on the projective C^*-tensor product $A \otimes_\pi B$. One aims at a characterization of this convex set by inequalities. While a complete characterization is unknown, a simple such inequality has been deduced from the work of Bell in the late 1950s on the Einstein–Podolsky–Rosen paradox. Therefore, inequalities of this type are often referred to as (generalized) *Bell's inequalities*. See also [9].

Lemma 8. *Let A and B be C^*-algebras and let τ be a decomposable state on the projective C^*-tensor product $A \otimes_\pi B$. Then*

$$|\tau(a \otimes (b - b'))| + |\tau(a' \otimes (b + b'))| \leq 2 \qquad (1.6)$$

holds for all self-adjoint elements $a, a' \in A$, $b, b' \in B$ of norm ≤ 1.

Proof. For a product state $\tau = \mu \otimes \nu$, we have

$$\tau(a \otimes (b - b')) = \mu(a) \cdot \nu(b) - \mu(a) \cdot \nu(b')$$
$$= \mu(a) \cdot \nu(b) \cdot (1 \pm \mu(a') \cdot \nu(b')) - \mu(a) \cdot \nu(b') \cdot (1 \pm \mu(a') \cdot \nu(b)).$$

By assumption, $|\mu(a)|, |\mu(a')|, |\nu(b)|, |\nu(b')| \leq 1$, so we have

$$|\tau(a \otimes (b - b'))| \leq |1 \pm \mu(a') \cdot \nu(b')| + |1 \pm \mu(a') \cdot \nu(b)|$$
$$= 1 \pm \mu(a') \cdot \nu(b') + 1 \pm \mu(a') \cdot \nu(b)$$
$$= 2 \pm \tau(a' \otimes (b + b')).$$

Hence, (1.6) holds for all product states. If τ is a convex combination of product states, $\tau = \sum_{j=1}^{n} \lambda_j \mu_j \otimes \nu_j$, we obtain

$$|\tau(a \otimes (b - b'))| + |\tau(a' \otimes (b + b'))|$$
$$\leq \sum_{j=1}^{n} \lambda_j \cdot \left\{ (\mu_j \otimes \nu_j)(a \otimes (b - b')) + (\mu_j \otimes \nu_j)(a' \otimes (b + b')) \right\}$$
$$\leq 2.$$

Taking pointwise limits of convex combinations, the inequality holds by continuity.   □

Example 15. Let $A = B = \mathrm{Mat}(2 \times 2; \mathbb{C})$ be matrix algebras. Let e_1, e_2 be the standard basis of $\mathbb{C}^2$. On $A \otimes B$, we have the Bell state τ, which is the vector state with the vector

$$\Omega := \frac{1}{\sqrt{2}} (e_1 \otimes e_1 + e_2 \otimes e_2).$$

It is easy to see that the Bell state is entangled. For instance, the observables

$$a = \begin{pmatrix} 1 & 0 \\ 0 & -1 \end{pmatrix}, \; a' = \begin{pmatrix} 0 & 1 \\ 1 & 0 \end{pmatrix}, \; b = \frac{1}{\sqrt{2}} \cdot \begin{pmatrix} 1 & 1 \\ 1 & -1 \end{pmatrix}, \; b' = \frac{1}{\sqrt{2}} \cdot \begin{pmatrix} -1 & 1 \\ 1 & 1 \end{pmatrix}$$

in the state τ yield

$$\tau(a \otimes (b - b')) = \sqrt{2} \, \tau(a \otimes a) = \sqrt{2} \, \langle (a \otimes a)(\Omega), \Omega \rangle$$
$$= \sqrt{2} \, \langle \Omega, \Omega \rangle = \sqrt{2}$$

and similarly

$$\tau(a' \otimes (b + b')) = \sqrt{2},$$

hence

$$|\tau(a \otimes (b - b'))| + |\tau(a' \otimes (b + b'))| = 2\sqrt{2} > 2.$$

Thus the state τ violates Bell's inequality and is therefore entangled by Lemma 8.

Bell's inequalities are often referred to as inequalities which a priori hold for all states in a classical system. The existence of entangled states may thus be considered as a characterizing phenomenon of quantum systems. In fact, if one of the observable algebras is abelian – e.g., if it corresponds to a classical system – then there are no entangled states in $A \otimes_\pi B$.

Proposition 6. *Let A and B be C^*-algebras with unit. If A or B is abelian, then all states on the projective C^*-tensor product $A \otimes_\pi B$ are decomposable.*

Proof. By Theorem 2, it suffices to show that every pure state τ on $A \otimes_\pi B$ is a product state. We first claim that $\tau(xy) = \tau(x) \cdot \tau(y)$ holds for all $x \in A \otimes_\pi B$ and all $y \in Z(A \otimes_\pi B)$, where $Z(A \otimes_\pi B)$ denotes the center of $A \otimes_\pi B$. Since $Z(A \otimes_\pi B)$ is spanned by its positive elements of norm ≤ 1, it suffices to prove the claim for y positive, i.e., $y = z^2$ for a self-adjoint $z \in Z(A \otimes_\pi B)$, with $\|y\| \leq 1$. If $\tau(y) = 0$, the Cauchy–Schwarz inequality

$$|\tau(xy)|^2 = |\tau((zx^*)^*z)|^2 \leq \tau(xz^*zx^*) \cdot \tau(z^*z)$$
$$= \tau(xyx^*) \cdot \tau(y)$$

implies $\tau(xy) = 0$. If $\tau(y) = 1$, then $\tau(1 - y) = 0$; thus $0 = \tau(x(1 - y)) = \tau(x) \cdot \tau(y) - \tau(xy)$.

For $0 < \tau(y) < 1$, we have

$$\tau(x) = \underbrace{\tau(y) \cdot \frac{1}{\tau(y)} \cdot \tau(xy)}_{=:\tau_1(x)} + \underbrace{(1 - \tau(y)) \cdot \frac{1}{1 - \tau(y)} \cdot \tau(x(1 - y))}_{=:\tau_2(x)} \quad \forall x \in A \otimes_\pi B\,.$$

Since $y \in Z(A \otimes_\pi B)$, we have $\tau_1(x^*x) = \frac{1}{\tau(y)} \cdot \tau(x^*xy) = \frac{1}{\tau(y)} \cdot \tau((zx)^*zx) \geq 0$. Similarly, $\tau_2(x^*x) = \frac{1}{1-\tau(y)} \cdot \tau(x^*x(1 - y)) \geq 0$, since

$$\tau(x^*xy) = \tau(x^*z^*zx) \leq \|z^*z\| \cdot \tau(x^*x) \leq \tau(x^*x)\,.$$

Clearly, $\tau_1(1) = \tau_2(1) = 1$; hence τ_1 and τ_2 are states on $A \otimes_\pi B$. Since τ is a pure state by assumption, we conclude $\tau = \tau_1 = \tau_2$. Hence $\tau_1(x) = \tau_2(x)$ for all $x \in A \otimes_\pi B$, which yields $\tau(xy) = \tau(y) \cdot \tau(x)$.

Now if A is abelian, then $A \otimes_\pi \{1\} \subset Z(A \otimes_\pi B)$. As we have seen, every pure state τ on $A \otimes_\pi B$ satisfies

$$\tau(a \otimes b) = \tau((a \otimes 1) \cdot (1 \otimes b)) = \tau^A(a) \cdot \tau^B(b) \quad \forall a \in A, b \in B\,.$$

Hence τ is a product state. $\qquad\qquad\qquad\qquad\qquad\qquad\qquad\qquad\qquad\qquad\qquad\qquad$ $\square$

1.6 Weyl Systems

In this section we introduce Weyl systems and CCR representations. They formal-
ize the "canonical commutator relations" from quantum field theory in an "expo-
nentiated form." The main result of this section is Theorem 3 which says that for
each symplectic vector space there is an essentially unique CCR representation.
Our approach follows ideas in [7]. A different proof of this result may be found in
[8, Sect. 5.2.2.2].

Let (V, ω) be a *symplectic vector space*, i.e., V is a real vector space of finite or
infinite dimension and $\omega : V \times V \to \mathbb{R}$ is an antisymmetric bilinear map such that
$\omega(\phi, \psi) = 0$ for all $\psi \in V$ implies $\phi = 0$.

Definition 22. *A Weyl system of* (V, ω) *consists of a C^*-algebra A with unit and a
map* $W : V \to A$ *such that for all* $\phi, \psi \in V$ *we have*

  (i) $W(0) = 1$,
  (ii) $W(-\phi) = W(\phi)^*$,
  (iii) $W(\phi) \cdot W(\psi) = e^{-i\omega(\phi,\psi)/2} W(\phi + \psi)$.

Condition (iii) says that W is a representation of the additive group V in A up to
the "twisting factor" $e^{-i\omega(\phi,\psi)/2}$. Note that since V is not given a topology there is no
requirement on W to be continuous. In fact, we will see that even in the case when
V is finite dimensional and so V carries a canonical topology W will in general not
be continuous.

Example 16. We construct a Weyl system for an arbitrary symplectic vector space
(V, ω). Let $H = L^2(V, \mathbb{C})$ be the Hilbert space of square-integrable complex-valued
functions on V with respect to the counting measure, i.e., H consists of those functions $F : V \to \mathbb{C}$ that vanish everywhere except for countably many points and
satisfy

$$\|F\|_{L^2}^2 := \sum_{\phi \in V} |F(\phi)|^2 < \infty.$$

The Hermitian product on H is given by

$$(F, G)_{L^2} = \sum_{\phi \in V} \overline{F(\phi)} \cdot G(\phi).$$

Let $A := \mathcal{L}(H)$ be the C^*-algebra of bounded linear operators on H as in Example 1. We define the map $W : V \to A$ by

$$(W(\phi)F)(\psi) := e^{i\omega(\phi,\psi)/2} F(\phi + \psi).$$

Obviously, $W(\phi)$ is a bounded linear operator on H for any $\phi \in V$ and $W(0) =
\mathrm{id}_H = 1$. We check (ii) by making the substitution $\chi = \phi + \psi$:

$$(W(\phi)F, G)_{L^2} = \sum_{\psi \in V} \overline{(W(\phi)F)(\psi)}\, G(\psi)$$

$$= \sum_{\psi \in V} \overline{e^{i\omega(\phi,\psi)/2}\, F(\phi + \psi)}\, G(\psi)$$

$$= \sum_{\chi \in V} \overline{e^{i\omega(\phi,\chi-\phi)/2}\, F(\chi)}\, G(\chi - \phi)$$

$$= \sum_{\chi \in V} \overline{e^{i\omega(\phi,\chi)/2}} \cdot \overline{F(\chi)} \cdot G(\chi - \phi)$$

$$= \sum_{\chi \in V} \overline{F(\chi)} \cdot e^{i\omega(-\phi,\chi)/2} \cdot G(\chi - \phi)$$

$$= (F, W(-\phi)G)_{L^2}.$$

Hence $W(\phi)^* = W(-\phi)$. To check (iii) we compute

$$(W(\phi)(W(\psi)F))(\chi) = e^{i\omega(\phi,\chi)/2}\,(W(\psi)F)(\phi + \chi)$$

$$= e^{i\omega(\phi,\chi)/2}\, e^{i\omega(\psi,\phi+\chi)/2}\, F(\phi + \chi + \psi)$$

$$= e^{i\omega(\psi,\phi)/2}\, e^{i\omega(\phi+\psi,\chi)/2}\, F(\phi + \chi + \psi)$$

$$= e^{-i\omega(\phi,\psi)/2}\,(W(\phi + \psi)F)(\chi).$$

Thus $W(\phi)W(\psi) = e^{-i\omega(\phi,\psi)/2}\, W(\phi + \psi)$. Let CCR$(V, \omega)$ be the C^*-subalgebra of $\mathcal{L}(H)$ generated by the elements $W(\phi)$, $\phi \in V$. Then CCR(V, ω) together with the map W forms a Weyl system for (V, ω).

Proposition 7. *Let (A, W) be a Weyl system of a symplectic vector space (V, ω). Then*

1. *$W(\phi)$ is unitary for each $\phi \in V$,*
2. *$\|W(\phi) - W(\psi)\| = 2$ for all $\phi, \psi \in V$, $\phi \neq \psi$,*
3. *the algebra A is not separable unless $V = \{0\}$,*
4. *the family $\{W(\phi)\}_{\phi \in V}$ is linearly independent.*

Proof. From $W(\phi)^*W(\phi) = W(-\phi)\,W(\phi) = e^{i\omega(-\phi,\phi)}W(0) = 1$ and similarly $W(\phi)\,W(\phi)^* = 1$ we see that $W(\phi)$ is unitary.

To show Assertion 2 let $\phi, \psi \in V$ with $\phi \neq \psi$. For arbitrary $\chi \in V$ we have

$$W(\chi)\,W(\phi - \psi)\,W(\chi)^{-1} = W(\chi)\,W(\phi - \psi)\,W(\chi)^*$$

$$= e^{-i\omega(\chi,\phi-\psi)/2}\, W(\chi + \phi - \psi)\,W(-\chi)$$

$$= e^{-i\omega(\chi,\phi-\psi)/2}\, e^{-i\omega(\chi+\phi-\psi,-\chi)/2}\, W(\chi + \phi - \psi - \chi)$$

$$= e^{-i\omega(\chi,\phi-\psi)}\, W(\phi - \psi).$$

Hence the spectrum satisfies

$$\sigma_A(W(\phi - \psi)) = \sigma_A(W(\chi)\,W(\phi - \psi)\,W(\chi)^{-1}) = e^{-i\omega(\chi,\phi-\psi)}\,\sigma_A(W(\phi - \psi)).$$

Since $\phi - \psi \neq 0$ the real number $\omega(\chi, \phi - \psi)$ runs through all of $\mathbb{R}$ as χ runs through V. Therefore the spectrum of $W(\phi - \psi)$ is U(1)-invariant. By Assertion 5 of Proposition 2 the spectrum is contained in S^1 and by Proposition 1 it is nonempty. Hence $\sigma_A(W(\phi - \psi)) = S^1$ and therefore

$$\sigma_A(e^{i\omega(\psi,\phi)/2} W(\phi - \psi)) = S^1.$$

Thus $\sigma_A(e^{i\omega(\psi,\phi)/2} W(\phi - \psi) - 1)$ is the circle of radius 1 centered at -1. Now Assertion 3 of Proposition 2 says

$$\|e^{i\omega(\psi,\phi)/2} W(\phi - \psi) - 1\| = \rho_A\left(e^{i\omega(\psi,\phi)/2} W(\phi - \psi) - 1\right) = 2.$$

From $W(\phi) - W(\psi) = W(\psi)(W(\psi)^* W(\phi) - 1) = W(\psi)(e^{i\omega(\psi,\phi)/2} W(\phi - \psi) - 1)$ we conclude

$$\begin{aligned}
\|W(\phi) &- W(\psi)\|^2 \\
&= \|(W(\phi) - W(\psi))^*(W(\phi) - W(\psi))\| \\
&= \|(e^{i\omega(\psi,\phi)/2} W(\phi - \psi) - 1)^* W(\psi)^* W(\psi)(e^{i\omega(\psi,\phi)/2} W(\phi - \psi) - 1)\| \\
&= \|(e^{i\omega(\psi,\phi)/2} W(\phi - \psi) - 1)^* (e^{i\omega(\psi,\phi)/2} W(\phi - \psi) - 1)\| \\
&= \|e^{i\omega(\psi,\phi)/2} W(\phi - \psi) - 1\|^2 \\
&= 4.
\end{aligned}$$

This shows part 2. Assertion 3 now follows directly since the balls of radius 1 centered at $W(\phi)$, $\phi \in V$ form an uncountable collection of mutually disjoint open subsets.

We show Assertion 4. Let $\phi_j \in V$, $j = 1, \ldots, n$ be pairwise different and let $\sum_{j=1}^n \alpha_j W(\phi_j) = 0$. We show $\alpha_1 = \cdots = \alpha_n = 0$ by induction on n. The case $n = 1$ is trivial by Assertion 1. Without loss of generality assume $\alpha_n \neq 0$. Hence

$$W(\phi_n) = \sum_{j=1}^{n-1} \frac{-\alpha_j}{\alpha_n} W(\phi_j)$$

and therefore

$$\begin{aligned}
1 &= W(\phi_n)^* W(\phi_n) \\
&= \sum_{j=1}^{n-1} \frac{-\alpha_j}{\alpha_n} W(-\phi_n) W(\phi_j) \\
&= \sum_{j=1}^{n-1} \frac{-\alpha_j}{\alpha_n} e^{-i\omega(-\phi_n,\phi_j)/2} W(\phi_j - \phi_n) \\
&= \sum_{j=1}^{n-1} \beta_j W(\phi_j - \phi_n),
\end{aligned}$$

where we have put $\beta_j := \frac{-\alpha_j}{\alpha_n} e^{i\omega(\phi_n,\phi_j)/2}$. For an arbitrary $\psi \in V$ we obtain

$$1 = W(\psi) \cdot 1 \cdot W(-\psi)$$

$$= \sum_{j=1}^{n-1} \beta_j W(\psi) W(\phi_j - \phi_n) W(-\psi)$$

$$= \sum_{j=1}^{n-1} \beta_j e^{-i\omega(\psi,\phi_j-\phi_n)} W(\phi_j - \phi_n).$$

From

$$\sum_{j=1}^{n-1} \beta_j W(\phi_j - \phi_n) = \sum_{j=1}^{n-1} \beta_j e^{-i\omega(\psi,\phi_j-\phi_n)} W(\phi_j - \phi_n)$$

we conclude by the induction hypothesis

$$\beta_j = \beta_j e^{-i\omega(\psi,\phi_j-\phi_n)}$$

for all $j = 1, \ldots, n - 1$. If some $\beta_j \neq 0$, then $e^{-i\omega(\psi,\phi_j-\phi_n)} = 1$, hence

$$\omega(\psi, \phi_j - \phi_n) = 0$$

for all $\psi \in V$. Since ω is nondegenerate $\phi_j - \phi_n = 0$, a contradiction. Therefore all β_j and thus all α_j are zero, a contradiction.   □

Remark 17. Let (A, W) be a Weyl system of the symplectic vector space (V, ω). Then the linear span of the $W(\phi)$, $\phi \in V$, is closed under multiplication and under $*$. This follows directly from the properties of a Weyl system. We denote this linear span by $\langle W(V) \rangle \subset A$. Now if (A', W') is another Weyl system of the same symplectic vector space (V, ω), then there is a unique linear map $\pi : \langle W(V) \rangle \to \langle W'(V) \rangle$ determined by $\pi(W(\phi)) = W'(\phi)$. Since π is given by a bijection on the bases $\{W(\phi)\}_{\phi \in V}$ and $\{W'(\phi)\}_{\phi \in V}$ it is a linear isomorphism. By the properties of a Weyl system π is a $*$-isomorphism. In other words, there is a unique $*$-isomorphism such that the following diagram commutes:

$$
\begin{array}{ccc}
 & & \langle W'(V) \rangle \\
 & \nearrow{\scriptstyle W_2} & \uparrow{\scriptstyle \pi} \\
V & \xrightarrow[\;W_1\;]{} & \langle W(V) \rangle
\end{array}
$$

Remark 18. On $\langle W(V) \rangle$ we can define the norm

$$\left\| \sum_\phi a_\phi W(\phi) \right\|_1 := \sum_\phi |a_\phi|.$$

This norm is not a C^*-norm but for every C^*-norm $\| \cdot \|_0$ on $\langle W(V) \rangle$ we have by the triangle inequality and by Assertion 1 of Proposition 7

$$\|a\|_0 \leq \|a\|_1 \tag{1.7}$$

for all $a \in \langle W(V) \rangle$.

Lemma 9. *Let (A, W) be a Weyl system of a symplectic vector space (V, ω). Then*

$$\|a\|_{max} := \sup\{\|a\|_0 \mid \| \cdot \|_0 \text{ is a } C^*\text{-norm on } \langle W(V) \rangle\}$$

defines a C^*-norm on $\langle W(V) \rangle$.

Proof. The given C^*-norm on A restricts to one on $\langle W(V) \rangle$, so the supremum is not taken on the empty set. Estimate (1.7) shows that the supremum is finite. The properties of a C^*-norm are easily checked, e.g., the triangle inequality follows from

$$
\begin{aligned}
\|a + b\|_{max} &= \sup\{\|a + b\|_0 \mid \| \cdot \|_0 \text{ is a } C^*\text{--norm on } \langle W(V) \rangle\} \\
&\leq \sup\{\|a\|_0 + \|b\|_0 \mid \| \cdot \|_0 \text{ is a } C^*\text{--norm on } \langle W(V) \rangle\} \\
&\leq \sup\{\|a\|_0 \mid \| \cdot \|_0 \text{ is a } C^*\text{--norm on } \langle W(V) \rangle\} \\
&\quad + \sup\{\|b\|_0 \mid \| \cdot \|_0 \text{ is a } C^*\text{--norm on } \langle W(V) \rangle\} \\
&= \|a\|_{max} + \|b\|_{max}.
\end{aligned}
$$

The other properties are shown similarly. $\square$

Lemma 10. *Let (A, W) be a Weyl system of a symplectic vector space (V, ω). Then the completion $\overline{\langle W(V) \rangle}^{max}$ of $\langle W(V) \rangle$ with respect to $\| \cdot \|_{max}$ is simple, i.e., it has no nontrivial closed two-sided $*$-ideals.*

Proof. By Remark 17 we may assume that (A, W) is the Weyl system constructed in Example 16. In particular, $\langle W(V) \rangle$ carries the C^*-norm $\| \cdot \|_{Op}$, the operator norm given by $\langle W(V) \rangle \subset \mathcal{L}(H)$ where $H = L^2(V, \mathbb{C})$.

Let $I \subset \overline{\langle W(V) \rangle}^{max}$ be a closed two-sided $*$-ideal. Then $I_0 := I \cap \mathbb{C} \cdot W(0)$ is a (complex) vector subspace in $\mathbb{C} \cdot W(0) = \mathbb{C} \cdot 1 \cong \mathbb{C}$ and thus $I_0 = \{0\}$ or $I_0 = \mathbb{C} \cdot W(0)$. If $I_0 = \mathbb{C} \cdot W(0)$, then I contains 1 and therefore $I = \overline{\langle W(V) \rangle}^{max}$. Hence we may assume $I_0 = \{0\}$.

Now we look at the projection map

$$P : \langle W(V) \rangle \to \mathbb{C} \cdot W(0), \quad P(\sum_\phi a_\phi W(\phi)) = a_0 W(0).$$

We check that P extends to a bounded operator on $\overline{\langle W(V) \rangle}^{max}$. Let $\delta_0 \in L^2(V, \mathbb{C})$ denote the function given by $\delta_0(0) = 1$ and $\delta_0(\phi) = 0$ otherwise. For $a = \sum_\phi a_\phi W(\phi)$ and $\psi \in V$ we have

$$(a \cdot \delta_0)(\psi) = (\sum_\phi a_\phi W(\phi)\delta_0)(\psi)$$

$$= \sum_\phi a_\phi \, e^{i\omega(\phi,\psi)/2} \delta_0(\phi + \psi)$$

$$= a_{-\psi} \, e^{i\omega(-\psi,\psi)/2} \quad = \quad a_{-\psi},$$

and therefore

$$(\delta_0, a \cdot \delta_0)_{L^2} = \sum_{\psi \in V} \overline{\delta_0(\psi)}(a \cdot \delta_0)(\psi) = (a \cdot \delta_0)(0) = a_0.$$

Moreover, $\|\delta_0\| = 1$. Thus

$$\|P(a)\|_{\max} = \|a_0 W(0)\|_{\max} = |a_0| = |(\delta_0, a \cdot \delta_0)_{L^2}| \leq \|a\|_{\mathrm{Op}} \leq \|a\|_{\max},$$

which shows that P extends to a bounded operator on $\overline{\langle W(V)\rangle}^{\max}$.

Now let $a \in I \subset \overline{\langle W(V)\rangle}^{\max}$. Fix $\epsilon > 0$. We write

$$a = a_0 W(0) + \sum_{j=1}^n a_j \, W(\phi_j) + r,$$

where the $\phi_j \neq 0$ are pairwise different and the remainder term r satisfies $\|r\|_{\max} < \epsilon$. For any $\psi \in V$ we have

$$I \ni W(\psi) a \, W(-\psi) = a_0 W(0) + \sum_{j=1}^n a_j \, e^{-i\omega(\psi,\phi_j)} \, W(\phi_j) + r(\psi),$$

where $\|r(\psi)\|_{\max} = \|W(\psi) r \, W(-\psi)\|_{\max} \leq \|r\|_{\max} < \epsilon$. If we choose ψ_1 and ψ_2 such that $e^{-i\omega(\psi_1,\phi_n)} = -e^{-i\omega(\psi_2,\phi_n)}$, then adding the two elements

$$a_0 W(0) + \sum_{j=1}^n a_j \, e^{-i\omega(\psi_1,\phi_j)} \, W(\phi_j) + r(\psi_1) \in I$$

$$a_0 W(0) + \sum_{j=1}^n a_j \, e^{-i\omega(\psi_2,\phi_j)} \, W(\phi_j) + r(\psi_2) \in I$$

yields

$$a_0 W(0) + \sum_{j=1}^{n-1} a_j' \, W(\phi_j) + r_1 \in I,$$

where $\|r_1\|_{\max} = \|\frac{r(\psi_1)+r(\psi_2)}{2}\|_{\max} < \frac{\epsilon+\epsilon}{2} = \epsilon$. Repeating this procedure we eventually get

$$a_0\, W(0) + r_n \in I,$$

where $\|r_n\|_{\max} < \epsilon$. Since ϵ is arbitrary and I is closed we conclude

$$P(a) = a_0\, W(0) \in I_0,$$

thus $a_0 = 0$.

For $a = \sum_\phi a_\phi\, W(\phi) \in I$ and arbitrary $\psi \in V$ we have $W(\psi)a \in I$ as well, hence $P(W(\psi)a) = 0$. This means $a_{-\psi} = 0$ for all ψ, thus $a = 0$. This shows $I = \{0\}$. $\qquad\qquad\square$

Definition 23. *A Weyl system (A, W) of a symplectic vector space (V, ω) is called a* CCR representation *of (V, ω) if A is generated as a C^*-algebra by the elements $W(\phi)$, $\phi \in V$. In this case we call A a* CCR-algebra *of (V, ω).*

Of course, for any Weyl system (A, W) we can simply replace A by the C^*-subalgebra generated by the elements $W(\phi)$, $\phi \in V$, and we obtain a CCR representation.

Existence of Weyl systems, and hence CCR representations, has been established in Example 16. Uniqueness also holds in the appropriate sense.

Theorem 3. *Let (V, ω) be a symplectic vector space and let (A_1, W_1) and (A_2, W_2) be two CCR representations of (V, ω).*

Then there exists a unique $*$-isomorphism $\pi : A_1 \to A_2$ such that the diagram

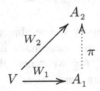

commutes.

Proof. We have to show that the $*$-isomorphism $\pi : \langle W_1(V)\rangle \to \langle W_2(V)\rangle$ as constructed in Remark 17 extends to an isometry $(A_1, \|\cdot\|_1) \to (A_2, \|\cdot\|_2)$. Since the pullback of the norm $\|\cdot\|_2$ on A_2 to $\langle W_1(V)\rangle$ via π is a C^*-norm we have $\|\pi(a)\|_2 \le \|a\|_{\max}$ for all $a \in \langle W_1(V)\rangle$. Hence π extends to a $*$–morphism $\overline{\langle W_1(V)\rangle}^{\max} \to A_2$. By Lemma 10 the kernel of π is trivial, hence π is injective. Proposition 4 implies that $\pi : (\overline{\langle W_1(V)\rangle}^{\max}, \|\cdot\|_{\max}) \to (A_2, \|\cdot\|_2)$ is an isometry.

In the special case $(A_1, \|\cdot\|_1) = (A_2, \|\cdot\|_2)$ where π is the identity this yields $\|\cdot\|_{\max} = \|\cdot\|_1$. Thus for arbitrary A_2 the map π extends to an isometry $(A_1, \|\cdot\|_1) \to (A_2, \|\cdot\|_2)$. $\qquad\square$

From now on we will call CCR(V, ω) as defined in Example 16 *the* CCR-algebra of (V, ω).

Corollary 5. *CCR-algebras of symplectic vector spaces are simple, i.e., all unit-preserving ∗-morphisms to other C^*-algebras are injective.*

Proof. Direct consequence of Corollary 2 and Lemma 10.   □

Corollary 6. *Let (V_1, ω_1) and (V_2, ω_2) be two symplectic vector spaces and let S : $V_1 \to V_2$ be a symplectic linear map, i.e., $\omega_2(S\phi, S\psi) = \omega_1(\phi, \psi)$ for all $\phi, \psi \in V_1$.*

Then there exists a unique injective ∗-morphism $\mathrm{CCR}(S)$: $\mathrm{CCR}(V_1, \omega_1) \to \mathrm{CCR}(V_2, \omega_2)$ such that the diagram

$$
\begin{array}{ccc}
V_1 & \xrightarrow{\quad\quad S \quad\quad} & V_2 \\
{\scriptstyle W_1}\big\downarrow & & \big\downarrow{\scriptstyle W_2} \\
\mathrm{CCR}(V_1, \omega_1) & \xrightarrow{\quad \mathrm{CCR}(S) \quad} & \mathrm{CCR}(V_2, \omega_2)
\end{array}
$$

commutes.

Proof. One immediately sees that $(\mathrm{CCR}(V_2, \omega_2), W_2 \circ S)$ is a Weyl system of (V_1, ω_1). Theorem 3 yields the result.   □

From uniqueness of the map $\mathrm{CCR}(S)$ we conclude that $\mathrm{CCR}(\mathrm{id}_V) = \mathrm{id}_{\mathrm{CCR}(V,\omega)}$ and $\mathrm{CCR}(S_2 \circ S_1) = \mathrm{CCR}(S_2) \circ \mathrm{CCR}(S_1)$. In other words, we have constructed a functor

$$\mathrm{CCR} : \mathsf{SymplVec} \to \mathsf{C^*Alg},$$

where SymplVec denotes the category whose objects are symplectic vector spaces and whose morphisms are symplectic linear maps, i.e., linear maps $A : (V_1, \omega_1) \to (V_2, \omega_2)$ with $A^*\omega_2 = \omega_1$. By C*Alg we denote the category whose objects are C^*-algebras and whose morphisms are *injective* unit-preserving ∗–morphisms. Observe that symplectic linear maps are automatically injective.

In the case $V_1 = V_2$ the induced ∗-automorphisms $\mathrm{CCR}(S)$ are called *Bogoliubov transformation* in the physics literature.

References

1. Bär, C., Ginoux, N., Pfäffle, F.: Wave equations on Lorentzian manifolds and quantization. EMS Publishing House, Zürich (2007)
2. Bratteli, O., Robinson, D.W.: Operator Algebras and Quantum Statistical Mechanics I. Springer, Berlin Heidelberg (2002)
3. Davidson, K.: C_-algebras by example. AMS, Providence (1997)
4. Dixmier, J.: Les C*-algèbres et leurs représentations, 2nd edition, Gauthier-Villars Éditeur, Paris (1969)
5. Murphy, G.: C*-algebras and operator theory. Academic Press, Boston (1990)

6. Takesaki, M.: Theory of Operator Algebra I, Springer, Berlin, Heidelberg, New York (2002)
7. Manuceau, J.: C*-algèbre de relations de commutation. Ann. Inst. H. Poincaré Sect. A (N.S.) **8**, 139 (1968)
8. Bratteli, O., Robinson, D.W.: Operator Algebras and Quantum Statistical Mechanics II. Springer, Berlin Heidelberg 2002
9. Baez, F.: Bell's inequality for C^*-Algebras. Lett. Math. Phys. **13**(2), 135–136 (1987)

Chapter 2
Lorentzian Manifolds

Frank Pfäffle

In this chapter some basic notions from Lorentzian geometry will be reviewed. In particular causality relations will be explained, Cauchy hypersurfaces and the concept of global hyperbolic manifolds will be introduced. Finally the structure of globally hyperbolic manifolds will be discussed.

More comprehensive introductions can be found in [1] and [2].

2.1 Preliminaries on Minkowski Space

Let V be an n-dimensional real vector space. A *Lorentzian scalar product* on V is a nondegenerate symmetric bilinear form $\langle\!\langle \cdot, \cdot \rangle\!\rangle$ of index 1. This means one can find a basis $e_1, \ldots, e_n$ of V such that

$$\langle\!\langle e_i, e_j \rangle\!\rangle = \begin{cases} -1 & \text{if } i = j = 1, \\ 1 & \text{if } i = j = 2, \ldots, n, \\ 0 & \text{otherwise.} \end{cases}$$

The simplest example for a Lorentzian scalar product on $\mathbb{R}^n$ is the Minkowski product $\langle\!\langle \cdot, \cdot \rangle\!\rangle_0$ given by $\langle\!\langle x, y \rangle\!\rangle_0 = -x_1 y_1 + x_2 y_2 + \cdots + x_n y_n$. In some sense this is the only example because from the above it follows that any n-dimensional vector space with Lorentzian scalar product $(V, \langle\!\langle \cdot, \cdot \rangle\!\rangle)$ is isometric to Minkowski space $(\mathbb{R}^n, \langle\!\langle \cdot, \cdot \rangle\!\rangle_0)$.

We denote the quadratic form associated with $\langle\!\langle \cdot, \cdot \rangle\!\rangle$ by

$$\gamma : V \to \mathbb{R}, \quad \gamma(X) := -\langle\!\langle X, X \rangle\!\rangle.$$

A vector $X \in V \setminus \{0\}$ is called *timelike* if $\gamma(X) > 0$, *lightlike* if $\gamma(X) = 0$ and $X \neq 0$, *causal* if timelike or lightlike, and *spacelike* if $\gamma(X) > 0$ or $X = 0$.

F. Pfäffle (✉)
Institut für Mathematik, Universität Potsdam, Am Neuen Palais 10, D-14469 Potsdam, Germany
e-mail: pfaeffle@math.uni-potsdam.de

Pfäffle, F.: *Lorentzian Manifolds*. Lect. Notes Phys. **786**, 39–58 (2009)
DOI 10.1007/978-3-642-02780-2_2   © Springer-Verlag Berlin Heidelberg 2009

Fig. 2.1 Lightcone in
Minkowski space

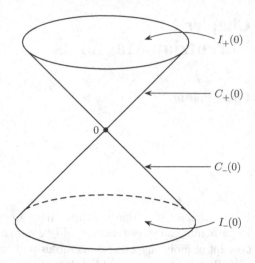

For $n \geq 2$ the set $I(0)$ of timelike vectors consists of two connected components. We choose a *time-orientation* on V by picking one of these two connected components. Denote this component by $I_+(0)$ and call its elements *future-directed*. We put $J_+(0) := \overline{I_+(0)}$, $C_+(0) := \partial I_+(0)$, $I_-(0) := -I_+(0)$, $J_-(0) := -J_+(0)$, and $C_-(0) := -C_+(0)$. Causal vectors in $J_+(0)$ (or in $J_-(0)$) are called *future-directed* (or *past-directed* respectively). (See Fig. 2.1.)

Remark 1. Given a positive number $\alpha > 0$ and a Lorentzian scalar product $\langle\!\langle \cdot, \cdot \rangle\!\rangle$ on a vector space V one gets another Lorentzian scalar product $\alpha \cdot \langle\!\langle \cdot, \cdot \rangle\!\rangle$. One observes that $X \in V$ is timelike with respect to $\langle\!\langle \cdot, \cdot \rangle\!\rangle$ if and only if it is timelike with respect to $\alpha \cdot \langle\!\langle \cdot, \cdot \rangle\!\rangle$. Analogously, the notion lightlike coincides for $\langle\!\langle \cdot, \cdot \rangle\!\rangle$ and $\alpha \cdot \langle\!\langle \cdot, \cdot \rangle\!\rangle$, and so do the notions causal and spacelike.

Hence, for both Lorentzian scalar products one gets the same set $I(0)$. If $\dim(V) \geq 2$ and we choose identical time-orientations for $\langle\!\langle \cdot, \cdot \rangle\!\rangle$ and $\alpha \cdot \langle\!\langle \cdot, \cdot \rangle\!\rangle$, the sets $I_\pm(0)$, $J_\pm(0)$, $C_\pm(0)$ are determined independently whether formed with respect to $\langle\!\langle \cdot, \cdot \rangle\!\rangle$ or $\alpha \cdot \langle\!\langle \cdot, \cdot \rangle\!\rangle$.

2.2 Lorentzian Manifolds

A *Lorentzian manifold* is a pair (M, g) where M is an n-dimensional smooth manifold and g is a Lorentzian metric, i.e., g associates with each point $p \in M$ a Lorentzian scalar product g_p on the tangent space T_pM.

One requires that g_p depends smoothly on p: This means that for any choice of local coordinates $x = (x_1, \ldots, x_n) : U \to V$, where $U \subset M$ and $V \subset \mathbb{R}^n$ are open subsets, and for any $i, j = 1, \ldots, n$ the functions $g_{ij} : V \to \mathbb{R}$ defined by $g_{ij} = g(\frac{\partial}{\partial x_i}, \frac{\partial}{\partial x_j})$ are smooth. Here $\frac{\partial}{\partial x_i}$ and $\frac{\partial}{\partial x_j}$ denote the coordinate vector fields as usual

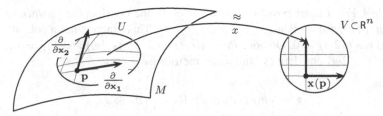

Fig. 2.2 Coordinate vectors $\frac{\partial}{\partial x_1}, \frac{\partial}{\partial x_2}$

(see Fig. 2.2). With respect to these coordinates one writes $g = \sum_{i,j} g_{ij} \, dx_i \otimes dx_j$ or shortly $g = \sum_{i,j} g_{ij} \, dx_i \, dx_j$.

Next we will give some prominent examples for Lorentzian manifolds.

Example 1. In cartesian coordinates $(x_1, \ldots, x_n)$ on $\mathbb{R}^n$ the *Minkowski metric* is defined by $g_{Mink} = -(dx_1)^2 + (dx_2)^2 + \cdots + (dx_n)^2$. This turns Minkowski space into a Lorentzian manifold.

Of course, the restriction of g_{Mink} to any open subset $U \subset \mathbb{R}^n$ yields a Lorentzian metric on U as well.

Example 2. Consider the unit circle $S^1 \subset \mathbb{R}^2$ with its standard metric $(d\theta)^2$. The *Lorentzian cylinder* is given by $M = S^1 \times \mathbb{R}$ together with the Lorentzian metric $g = -(d\theta)^2 + (dx)^2$.

Example 3. Let (N, h) be a connected Riemannian manifold and $I \subset \mathbb{R}$ an open interval. For any $t \in I$, $p \in N$ one identifies $T_{(t,p)}(I \times N) = T_t I \oplus T_p N$. Then for any smooth positive function $f : I \to (0, \infty)$ the Lorentzian metric $g = -dt^2 + f(t)^2 \cdot h$ on $I \times M$ is defined as follows: For any $\xi_1, \xi_2 \in T_{(t,p)}(I \times N)$ one writes $\xi_i = \left(\alpha_i \frac{d}{dt}\right) \oplus \zeta_i$ with $\alpha_i \in \mathbb{R}$ and $\zeta_i \in T_p N$, $i = 1, 2$, and one has $g(\xi_1, \xi_2) = -\alpha_1 \cdot \alpha_2 + f(t)^2 \cdot h(\zeta_1, \zeta_2)$. Such a Lorentzian metric g is called a *warped product metric* (Fig. 2.3).

This example covers *Robertson–Walker spacetimes* where one requires additionally that (N, h) is complete and has constant curvature. In particular *Friedmann cosmological models* are of this type. In general relativity they are used to discuss big bang, expansion of the universe, and cosmological redshift; compare [2, Chaps. 5 and 6] or [1, Chap. 12]. A special case of this is *deSitter spacetime* where $I = \mathbb{R}$, $N = S^{n-1}$, h is the canonical metric of S^{n-1} of constant sectional curvature 1, and $f(t) = \cosh(t)$.

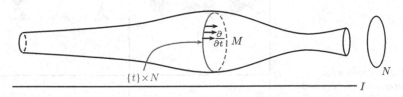

Fig. 2.3 Warped product

Example 4. For a fixed positive number $m > 0$ one considers the *Schwarzschild function* $h : (0, \infty) \to \mathbb{R}$ given by $h(r) = 1 - \frac{2m}{r}$. This function has a pole at $r = 0$ and one has $h(2m) = 0$. On both $P_I = \{(r, t) \in \mathbb{R}^2 \,|\, r > 2m\}$ and $P_{II} = \{(r, t) \in \mathbb{R}^2 \,|\, 0 < r < 2m\}$ one defines Lorentzian metrics by

$$g = -h(r) \cdot dt \otimes dt + \frac{1}{h(r)} \cdot dr \otimes dr,$$

and one calls (P_I, g) *Schwarzschild half-plane* and (P_{II}, g) *Schwarzschild strip*. For a tangent vector $\alpha \frac{\partial}{\partial t} + \beta \frac{\partial}{\partial r}$ being timelike is equivalent to $\alpha^2 > \frac{1}{h(r)^2} \beta^2$. Hence, one can illustrate the set of timelike vector in the tangent spaces $T_{(r,t)} P_I$, resp., $T_{(r,t)} P_{II}$ as in Fig. 2.4.

The "singularity" of the Lorentzian metric g for $r = 2m$ is not as crucial as it might seem at first glance, by a change of coordinates one can overcome this singularity (e.g., in the so-called *Kruskal coordinates*).

One uses (P_I, g) and (P_{II}, g) to discuss the exterior and the interior of a static rotationally symmetric black hole with mass m, compare [1, Chap. 13]. For this one considers the two-dimensional sphere S^2 with its natural Riemannian metric can_{S^2}, and on both $N = P_I \times S^2$ and $B = P_{II} \times S^2$ one gets a Lorentzian metric by

$$-h(r) \cdot dt \otimes dt + \frac{1}{h(r)} \cdot dr \otimes dr + r^2 \cdot \mathrm{can}_{S^2}.$$

Equipped with this metric, N is called *Schwarzschild exterior spacetime* and B *Schwarzschild black hole*, both of mass m.

Example 5. Let $S^{n-1} = \{(x_1, \ldots, x_n) \in \mathbb{R}^n \,|\, (x_1)^2 + \cdots + (x_n)^2 = 1\}$ be the n-dimensional sphere equipped with its natural Riemannian metric $\mathrm{can}_{S^{n-1}}$. The restriction of this metric to $S_+^{n-1} = \{(x_1, \ldots, x_n) \in S^{n-1} \,|\, x_n > 0\}$ is denoted by $\mathrm{can}_{S_+^{n-1}}$. Then, on $\mathbb{R} \times S_+^{n-1}$ one defines a Lorentzian metric by

$$g_{AdS} = \frac{1}{(x_n)^2} \cdot \left(-dt^2 + \mathrm{can}_{S_+^{n-1}} \right),$$

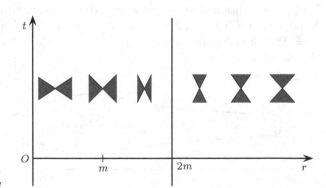

Fig. 2.4 Lightcones for Schwarzschild strip P_{II} and Schwarzschild half-plane P_I

and one calls $(\mathbb{R} \times S_+^{n-1}, g_{AdS})$ the n-dimensional *anti-deSitter spacetime*. This definition is not exactly the one given in [1, Chap. 8, p. 228f.], but one can show that both definitions coincide; compare [3, Chap. 3.5., p. 95ff.].

By Remark 1 we see that a tangent vector of $\mathbb{R} \times S_+^{n-1}$ is timelike (lightlike, spacelike) with respect to g_{AdS} if and only if it is so with respect to the Lorentzian metric $-dt^2 + \text{can}_{S_+^{n-1}}$.

In general relativity one is interested in four-dimensional anti-deSitter spacetime because it provides a vacuum solution of Einstein's field equation with cosmological constant $\Lambda = -3$; see [1, Chap. 14, Example 41].

2.3 Time-Orientation and Causality Relations

Let (M, g) denote a Lorentzian manifold of dimension $n \geq 2$. Then at each point $p \in M$ the set of timelike vectors in the tangent space $T_p M$ consists of two connected components, which cannot be distinguished intrinsically. A *time-orientation* on M is a choice $I_+(0) \subset T_p M$ of one of these connected components which depends continuously on p.

A time-orientation (Fig. 2.5) is given by a continuous timelike vector field τ on M which takes values in these chosen connected components: $\tau(p) \in I_+(0) \subset T_p M$ for each $p \in M$.

Definition 1. *One calls a Lorentzian manifold (M, g) time-orientable if there exists a continuous timelike vector field τ on M. A Lorentzian manifold (M, g) together with such a vector field τ is called* time-oriented. *In what follows time-oriented connected Lorentzian manifolds will be referred to as* spacetimes.

It should be noted that in contrast to the notion of orientability which only depends on the topology of the underlying manifold the concept of time-orientability depends on the Lorentzian metric.

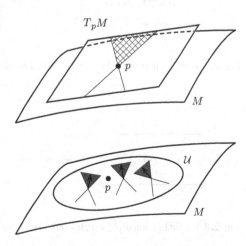

Fig. 2.5 Time-orientation

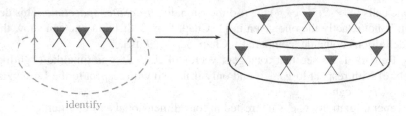

identify

Fig. 2.6 Example for orientable and time-orientable manifold

If we go through the list of examples from Sect. 2.2, we see that all these Lorentzian manifolds are time-orientable (Fig. 2.7). Timelike vector fields can be given as follows: on Minkowski space by $\frac{\partial}{\partial x_1}$, on the Lorentzian cylinder by $\frac{\partial}{\partial \theta}$, on the warped product in Example 3 by $\frac{d}{dt}$, on Schwarzschild exterior spacetime by $\frac{\partial}{\partial t}$, on Schwarzschild black hole by $\frac{\partial}{\partial r}$, and finally on anti-deSitter spacetime by $\frac{\partial}{\partial t}$.

From now on let (M, g) denote a spacetime of dimension $n \geq 2$. Then for each point $p \in M$ the tangent space $T_p M$ is a vector space equipped with the Lorentzian scalar product g_p and the time-orientation induced by the lightlike vector $\tau(p)$, and in $(T_p M, g_p)$ the notions of timelike, lightlike, causal, spacelike, future-directed vectors are defined as explained in Sect. 2.1.

Definition 2. *A continuous piecewise C^1-curve in M is called* timelike, lightlike, causal, spacelike, future-directed, *or* past-directed *if all its tangent vectors are timelike, lightlike, causal, spacelike, future-directed, or past-directed, respectively.*

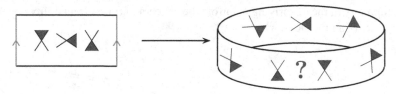

Fig. 2.7 Lorentzian manifold which is orientable, but not time-orientable

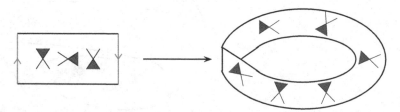

Fig. 2.8 Lorentzian manifold which is not orientable, but time-orientable

The *causality relations* on M are defined as follows: Let $p, q \in M$, then one has

$$p \ll q :\Longleftrightarrow \text{there is a future-directed timelike curve in } M \text{ from } p \text{ to } q,$$
$$p < q :\Longleftrightarrow \text{there is a future-directed causal curve in } M \text{ from } p \text{ to } q,$$
$$p \leq q :\Longleftrightarrow p < q \text{ or } p = q.$$

These causality relations are transitive as two causal (timelike) curves in M, say the one from p_1 to p_2 and the other from p_2 to p_3, can be put together to a piecewise causal (timelike) C^1-curve from p_1 to p_3.

Definition 3. *The* chronological future $I_+^M(x)$ *of a point* $x \in M$ *is the set of points that can be reached from x by future-directed timelike curves, i.e.,*

$$I_+^M(x) = \{y \in M \mid x < y\}.$$

Similarly, the causal future $J_+^M(x)$ *of a point* $x \in M$ *consists of those points that can be reached from x by future-directed causal curves and of x itself:*

$$J_+^M(x) = \{y \in M \mid x \leq y\}.$$

The chronological future *of a subset* $A \subset M$ *is defined to be*

$$I_+^M(A) := \bigcup_{x \in A} I_+^M(x).$$

Similarly, the causal future *of A is*

$$J_+^M(A) := \bigcup_{x \in A} J_+^M(x).$$

The chronological past $I_-^M(x)$ *resp.* $I_-^M(A)$ *and the* causal past $J_-^M(x)$ *resp.* $J_-^M(A)$ *are defined by replacing future-directed curves by past-directed curves.*

For $A \subset M$ one also uses the notation

$$J^M(A) = J_+^M(A) \cup J_-^M(A).$$

Remark 2. Evidently, for any $A \subset M$ one gets the inclusion $A \cup I_+^M(A) \subset J_+^M(A)$.

Example 6. We consider Minkowski space $(\mathbb{R}^2, g_{Mink})$. Then for $p \in \mathbb{R}^2$ the chronological future $I_+^{\mathbb{R}^2}(p) \subset \mathbb{R}^2$ is an open subset, and for a compact subset A of the x_2-axis the causal past $J_-^{\mathbb{R}^2}(A) \subset \mathbb{R}^2$ is a closed subset, as indicated in Fig. 2.9.

Example 7. By Example 1 every open subset of Minkowski space forms a Lorentzian manifold. Let M be two-dimensional Minkowski space with one point removed. Then there are subsets $A \subset M$ whose causal past is not closed as one can see in Fig. 2.10.

Fig. 2.9 Chronological future $I_+^{\mathbb{R}^2}(p)$ and causal past $J_-^{\mathbb{R}^2}(A)$

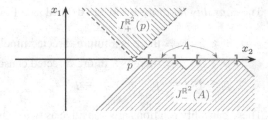

Fig. 2.10 Causal future and past of subset A of M

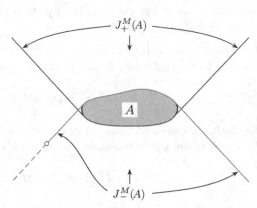

Example 8. If one unwraps Lorentzian cylinder $(M, g) = (S^1 \times \mathbb{R}, -d\theta^2 + dx_1{}^2)$ one can think of M as a strip in Minkowski space $\mathbb{R}^2$ for which the upper and lower boundaries are identified. In this picture it can easily be seen that $I_+^M(p) = J_+^M(p) = I_-^M(p) = J_-^M(p) = M$ for any $p \in M$; see Fig. 2.11.

Any connected open subset Ω of a spacetime M is a spacetime in its own right if one restricts the Lorentzian metric of M to Ω. Therefore $J_+^\Omega(x)$ and $J_-^\Omega(x)$ are well defined for $x \in \Omega$.

Definition 4. *A domain $\Omega \subset M$ in a spacetime is called* causally compatible *if for all points $x \in \Omega$ one has*

$$J_\pm^\Omega(x) = J_\pm^M(x) \cap \Omega.$$

Note that the inclusion "⊂" always holds. The condition of being causally compatible means that whenever two points in Ω can be joined by a causal curve in M this can also be done inside Ω (Fig. 2.12).

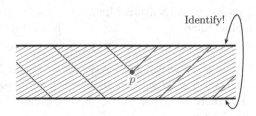

Fig. 2.11 $J_+^M(p) = M$

Fig. 2.12 Causally compatible subset of Minkowski space

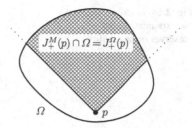

Fig. 2.13 Domain which is not causally compatible in Minkowski space

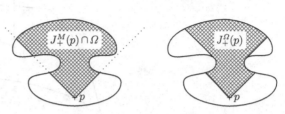

If $\Omega \subset M$ is a causally compatible domain in a spacetime, then we immediately see that for each subset $A \subset \Omega$ we have

$$J_\pm^\Omega(A) = J_\pm^M(A) \cap \Omega.$$

Note also that being causally compatible is transitive: If $\Omega \subset \Omega' \subset \Omega''$, if Ω is causally compatible in Ω', and if Ω' is causally compatible in Ω'', then so is Ω in Ω''.

Next, we recall the definition of the exponential map: For $p \in M$ and $\xi \in T_p M$ let c_ξ denote the (unique) geodesic with initial conditions $c_\xi(0) = p$ and $\dot{c}_\xi(0) = \xi$. One considers the set

$$\mathcal{D}_p = \left\{ \xi \in T_p M \,\middle|\, c_\xi \text{ can be defined at least on } [0, 1] \right\} \subset T_p M$$

and defines the *exponential map* $\exp_p : \mathcal{D}_p \to M$ by $\exp_p(\xi) = c_\xi(1)$.

One important feature of the exponential map is that it is an isometry in radial direction which is the statement of the following lemma.

Lemma 1 (Gauss Lemma). *Let $\xi \in \mathcal{D}_p$ and $\zeta_1, \zeta_2 \in T_\xi(T_p M) = T_p M$ with ζ_1 radial, i.e., there exists $t_0 \in \mathbb{R}$ with $\zeta_1 = t_0 \xi$, then*

$$g_{\exp_p(\xi)} \left(d\exp_p \big|_\xi (\zeta_1), d\exp_p \big|_\xi (\zeta_2) \right) = g_p(\zeta_1, \zeta_2).$$

A proof of the Gauss lemma can be found, e.g., in [1, Chap. 5, p. 126f.].

Lemma 2. *Let $p \in M$ and $b > 0$ and let $\widetilde{c} : [0, b] \longrightarrow T_p M$ be a piecewise smooth curve with $\widetilde{c}(0) = 0$ and $\widetilde{c}(t) \in \mathcal{D}_p$ for any $t \in [0, b]$. Suppose that $c := \exp_p \circ \widetilde{c} : [0, b] \longrightarrow M$ is a timelike future-directed curve, then*

$$\widetilde{c}(t) \in I_+(0) \subset T_p M \quad \text{for any } t \in [0, b].$$

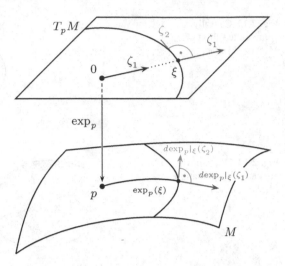

Fig. 2.14 In radial direction the exponential map preserves orthogonality

Proof. Suppose in addition that $\widetilde{c}$ is smooth. On T_pM we consider the quadratic form induced by the Lorentzian scalar product $\gamma : T_pM \longrightarrow \mathbb{R}$, $\gamma(\xi) = -g_p(\xi, \xi)$, and we compute $\operatorname{grad}\gamma(\xi) = -2\xi$. The Gauss lemma applied for $\xi \in \mathcal{D}_p$ and $\zeta_1 = \zeta_2 = 2\xi$ yields

$$g_{\exp_p(\xi)}\left(d\exp_p\big|_\xi(\zeta_1), d\exp_p\big|_\xi(\zeta_2)\right) = g_p(\zeta_1, \zeta_2) = -4\gamma(\xi).$$

Denote $P(\xi) = d\exp_p\big|_\xi(2\xi)$. Then by the above formula $P(\xi)$ is timelike whenever ξ is timelike.

From $\widetilde{c}(0) = 0$ and $(d/dt)\widetilde{c}(0) = d\exp_p\big|_0((d/dt)\widetilde{c}(0)) = \dot{c}(0) \in I_+(0)$ we get for a sufficiently small $\varepsilon > 0$ that $\widetilde{c}(t) \in I_+(0)$ for all $t \in (0, \varepsilon)$. Hence $P(\widetilde{c}(t))$ is timelike and future-directed for $t \in (0, \varepsilon)$.

For $\xi = \widetilde{c}(t)$, $\zeta_1 = 2\xi = -\operatorname{grad}\gamma(\xi)$ and $\zeta_2 = (d/dt)\widetilde{c}(t)$ the Gauss lemma gives

$$\frac{d}{dt}(\gamma \circ \widetilde{c})(t) = -g_p(\zeta_1, \zeta_2) = -g_{\exp_p(\xi)}\left(P(\widetilde{c}(t)), \dot{c}(t)\right).$$

If there were $t_1 \in (0, b]$ with $\gamma(\widetilde{c}(t_1)) = 0$, w.l.o.g. let t_1 be the smallest value in $(0, b]$ with $\gamma(\widetilde{c}(t_1)) = 0$, then one could find a $t_0 \in (0, t_1)$ with

$$0 = \frac{d}{dt}(\gamma \circ \widetilde{c})(t_0) = -g_{\exp_p(\xi)}\left(P(\widetilde{c}(t_0)), \dot{c}(t_0)\right).$$

On the other hand, having chosen t_1 minimally implies that $P(\widetilde{c}(t_0))$ is timelike and future-directed. Together with $\dot{c}(t_0) \in I_+^M(c(t_0))$ this yields $g_{\exp_p(\xi)}\left(P(\widetilde{c}(t_0)), \dot{c}(t_0)\right) < 0$, a contradiction.

Hence one has $\gamma(\widetilde{c}(t)) > 0$ for any $t \in (0, b]$, and the continuous curve $\widetilde{c}|_{(0,b]}$ does not leave the connected component of $I(0)$ in which it runs initially. This finishes the proof if one supposes that $\tilde{c}$ is smooth. For the proof in the general case see [1, Chap. 5, Lemma 33]. ☐

Definition 5. *A domain $\Omega \subset M$ is called* geodesically starshaped *with respect to a fixed point $p \in \Omega$ if there exists an open subset $\Omega' \subset T_pM$, starshaped with respect to 0, such that the Riemannian exponential map $\exp_x$ maps Ω' diffeomorphically onto Ω.*

One calls a domain $\Omega \subset M$ geodesically convex (or simply convex) if it is geodesically starshaped with respect to all of its points.

Remark 3. Every point of a Lorentzian manifold (which need not necessarily be a spacetime) possesses a convex neighborhood, see [1, Chap. 5, Prop. 7]. Furthermore, for each open covering of a Lorentzian manifold one can find a refinement consisting of convex open subsets, see [1, Chap. 5, Lemma 10].

Sometimes sets that are geodesically starshaped with respect to a point p are useful to get relations between objects defined in the tangent T_pM and objects defined on M. For the moment this will be illustrated by the following lemma.

Lemma 3. *Let M a spacetime and $p \in M$. Let the domain $\Omega \subset M$ be a geodesically starshaped with respect to p (Fig. 2.15). Let Ω' be an open neighborhood of 0 in T_pM such that Ω' is starshaped with respect to 0 and $\exp_p |_{\Omega'} : \Omega' \longrightarrow \Omega$ is a diffeomorphism. Then one has*

$$I_\pm^\Omega(p) = \exp_p\left(I_\pm(0) \cap \Omega'\right) \quad \text{and}$$
$$J_\pm^\Omega(p) = \exp_p\left(J_\pm(0) \cap \Omega'\right).$$

Proof. We will only prove the equation $I_+^\Omega(p) = \exp_p\left(I_+(0) \cap \Omega'\right)$.

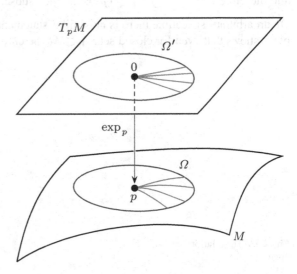

Fig. 2.15 Ω is geodesically starshaped with respect to p

For $q \in I_+^\Omega(p)$ one can find a future-directed timelike curve $c : [0, b] \to \Omega$ from p to q. We define the curve $\tilde{c} : [0, b] \to \Omega' \subset T_p M$ by $\tilde{c} = \exp_p^{-1} \circ c$ and get from Lemma 2 that $\tilde{c}(t) \in I_+(0)$ for $0 < t \le b$, in particular $\exp_p^{-1}(q) = \tilde{c}(b) \in I_+(0)$. This shows the inclusion $I_+^\Omega(p) \subset \exp_p \left(I_+(0) \cap \Omega' \right)$.

For the other inclusion we consider $\xi \in I_+(0) \cap \Omega'$. Then the map $t \mapsto t \cdot \xi$ takes its values in $I_+(0) \cap \Omega'$ as $t \in [0, 1]$. Therefore $\exp_p(t\xi)$ gives a timelike future-directed geodesic which stays in Ω as $t \in [0, 1]$, and it follows that $\exp_p(\xi) \in I_+^\Omega(p)$.

For a proof of $J_\pm^\Omega(p) = \exp_p \left(J_\pm(0) \cap \Omega' \right)$ we refer to [1, Chap. 14, Lemma 2]. $\qquad \square$

For Ω and Ω' as in Lemma 3 we put $C_\pm^\Omega(p) = \exp_p(C_\pm(0) \cap \Omega')$.

Proposition 1. *On any spacetime M the relation "$\ll$" is open, this means that for every $p, q \in M$ with $p \ll q$ there are open neighborhoods U and V of p and q, respectively, such that for any $p' \in U$ and $q' \in V$ one has $p' \ll q'$ (Fig. 2.16).*

Proof. For $p, q \in M$ with $p \ll q$ there are geodesically convex neighborhoods $\tilde{U}$, $\tilde{V}$, respectively. We can find a future-directed timelike curve c from p to q. Then we choose $\tilde{p} \in \tilde{U}$ and $\tilde{q} \in \tilde{V}$ sitting on c such that $p \ll \tilde{p} \ll \tilde{q} \ll q$. As $\tilde{U}$ is starshaped with respect to $\tilde{p}$ there is a starshaped open neighborhood $\tilde{\Omega}$ of 0 in $T_{\tilde{p}} M$ such that $\exp_{\tilde{p}} : \tilde{\Omega} \to \tilde{U}$ is a diffeomorphism. We set $U = I_-^{\tilde{U}}(\tilde{p})$, and Lemma 3 shows that $U = \exp_{\tilde{p}}(I_-(0) \cap \tilde{\Omega})$ is an open neighborhood of p in M. Analogously, one finds that $V = I_+^{\tilde{V}}(\tilde{q})$ is an open neighborhood of q. Finally, for any $p' \in U$ and $q' \in V$ one gets $p' \ll \tilde{p} \ll \tilde{q} \ll q'$ and hence $p' \ll q'$. $\qquad \square$

Corollary 1. *For an arbitrary $A \subset M$ the chronological future $I_+^M(A)$ and the chronological past $I_-^M(A)$ are open subsets in M.*
Proof. Proposition 1 implies that for any $p \in M$ the subset $I_+^M(p) \subset M$ is open, and therefore $I_+^M(A) = \bigcup_{p \in A} I_+^M(p)$ is an open subset of M as well. $\qquad \square$

On an arbitrary spacetime there is no similar statement for the relation "$\le$." Example 7 shows that even for closed sets $A \subset M$ the chronological future and past are

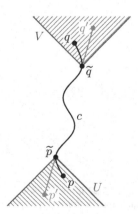

Fig. 2.16 The relation "$\ll$" is open

not always closed. In general one only has that $I_\pm^M(A)$ is the interior of $J_\pm^M(A)$ and that $J_\pm^M(A)$ is contained in the closure of $I_\pm^M(A)$.

Definition 6. *A domain Ω is called* causal *if its closure $\overline{\Omega}$ is contained in a convex domain Ω' and if for any $p, q \in \overline{\Omega}$ the intersection $J_+^{\Omega'}(p) \cap J_-^{\Omega'}(q)$ is compact and contained in $\overline{\Omega}$.*

Causal domains appear in the theory of wave equations: The local construction of fundamental solutions is always possible on causal domains provided their volume is small enough, see Proposition 3 on page 71.

Remark 4. Any point $p \in M$ in a spacetime possesses a causal neighborhood, compare [4, Theorem 4.4.1], and given a neighborhood $\widetilde{\Omega}$ of p, one can always find a causal domain Ω with $p \in \Omega \subset \widetilde{\Omega}$ (Fig. 2.17).

The last notion introduced in this section is needed if it comes to the discussion of uniqueness of solutions for wave equations:

Definition 7. *A subset $A \subset M$ is called* past-compact *if $A \cap J_-^M(p)$ is compact for all $p \in M$. Similarly, one defines* future-compact *subsets (Fig. 2.18).*

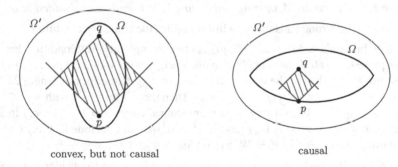

convex, but not causal   causal

Fig. 2.17 Convexity versus causality

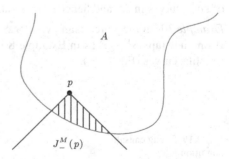

Fig. 2.18 The subset A is past-compact

2.4 Causality Condition and Global Hyperbolicity

In general relativity worldlines of particles are modeled by causal curves. If now the spacetime is compact something strange happens.

Proposition 2. *If the spacetime M is compact, there exists a closed timelike curve in M.*

Proof. The family $\{I_+^M(p)\}_{p \in M}$ is an open covering of M. By compactness one has $M = I_+^M(p_1) \cup \cdots \cup I_+^M(p_k)$ for suitably chosen $p_1, \ldots, p_k \in M$. We can assume that $p_1 \notin I_+^M(p_2) \cup \cdots \cup I_+^M(p_k)$, otherwise $p_1 \in I_+^M(p_m)$ for an $m \geq 2$ and hence $I_+^M(p_1) \subset I_+^M(p_m)$ and we can omit $I_+^M(p_1)$ in the finite covering. Therefore we can assume $p_1 \in I_+^M(p_1)$, and there is a timelike future-directed curve starting and ending in p_1.   □

In spacetimes with timelike loops one can produce paradoxes as travels into the past (like in science fiction). Therefore one excludes compact spacetimes, for physically reasonable spacetimes one requires the causality condition or the strong causality condition (Fig. 2.19).

Definition 8. *A spacetime is said to satisfy the* causality condition *if it does not contain any closed causal curve.*

A spacetime M is said to satisfy the strong causality condition *if there are no almost closed causal curves. More precisely, for each point $p \in M$ and for each open neighborhood U of p there exists an open neighborhood $V \subset U$ of p such that each causal curve in M starting and ending in V is entirely contained in U.*

Obviously, the strong causality condition implies the causality condition.

Example 9. In Minkowski space $(\mathbb{R}^n, g_{Mink})$ the strong causality condition holds. One can prove this as follows: Let U be an open neighborhood of $p = (p_1, \ldots, p_n) \in \mathbb{R}^n$. For any $\delta > 0$ denote the open cube with center p and edges of length 2δ by $W_\delta = (p_1 - \delta, p_1 + \delta) \times \cdots \times (p_n - \delta, p_n + \delta)$. Then there is an $\varepsilon > 0$ with $W_{2\varepsilon} \subset U$, and one can put $V = W_\varepsilon$. Observing that any causal curve $c = (c_1, \ldots, c_n)$ in $\mathbb{R}^n$ satisfies $(\dot{c}_1)^2 \geq (\dot{c}_2)^2 + \cdots + (\dot{c}_n)^2$ and $(\dot{c}_1)^2 > 0$, we can conclude that any causal curve starting and ending in $V = W_\varepsilon$ cannot leave $W_{2\varepsilon} \subset U$.

Remark 5. Let M satisfy the (strong) causality condition and consider any open connected subset $\Omega \subset M$ with induced Lorentzian metric as a spacetime. Then non-existence of (almost) closed causal curves in M directly implies non-existence of such curves in Ω, and hence also Ω satisfies the (strong) causality condition.

Example 10. In the Lorentzian cylinder $S^1 \times \mathbb{R}$ the causality condition is violated. If one unwraps $S^1 \times \mathbb{R}$ as in Example 8 it can be easily seen that there are closed timelike curves (Fig. 2.20).

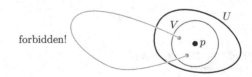

forbidden!

Fig. 2.19 Strong causality condition

Fig. 2.20 Closed timelike
curve c in Lorentzian cylinder

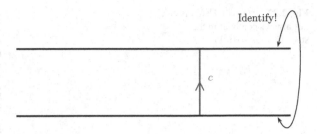

Example 11. Consider the spacetime M which is obtained from the Lorentzian
cylinder by removing two spacelike half-lines G_1 and G_2 whose endpoints can
be joined by a short lightlike curve, as indicated in Fig. 2.21. Then the causality
condition holds for M, but the strong causality condition is violated: For any p on
the short lightlike curve and any arbitrarily small neighborhood of p there is a causal
curve which starts and ends in this neighborhood but is not entirely contained.

Definition 9. *A spacetime M is called a* globally hyperbolic manifold *if it satisfies
the strong causality condition and if for all $p, q \in M$ the intersection $J_+^M(p) \cap J_-^M(q)$
is compact.*

The notion of global hyperbolicity has been introduced by J. Leray in [5]. Globally hyperbolic manifolds are interesting because they form a large class of spacetimes on which wave equations possess a very satisfying global solution theory; see
Chap. 3.

Example 12. In Minkowski space $(\mathbb{R}^n, g_{Mink})$ for any $p, q \in \mathbb{R}^n$ both $J_+^{\mathbb{R}^n}(p)$
and $J_-^{\mathbb{R}^n}(q)$ are closed. Furthermore $J_+^{\mathbb{R}^n}(p) \cap J_-^{\mathbb{R}^n}(q)$ is bounded (with respect to
Euclidean norm), and hence compact. In Example 9 we have already seen that for
$(\mathbb{R}^n, g_{Mink})$ the strong causality condition holds. Hence, Minkowski space is globally hyperbolic.

Example 13. As seen before, the Lorentzian cylinder $M = S^1 \times \mathbb{R}$ does not fulfill the
strong causality condition and is therefore not globally hyperbolic. Furthermore the
compactness condition in Definition 9 is violated because one has $J_+^M(p) \cap J_-^M(q) =
M$ for any $p, q \in M$.

Example 14. Consider the subset $\Omega = \mathbb{R} \times (0, 1)$ of two-dimensional Minkowski
space $(\mathbb{R}^2, g_{Mink})$. By Remark 5 the strong causality condition holds for Ω, but there

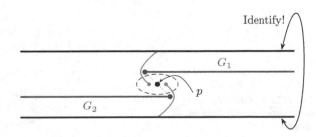

Fig. 2.21 Causality condition
holds but strong causality
condition is violated

Fig. 2.22 $J_+^\Omega(p) \cap J_-^\Omega(q)$ is
not always compact in the
strip $\Omega = \mathbb{R} \times (0, 1)$

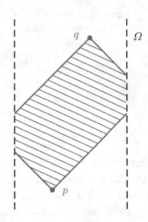

are points $p, q \in \Omega$ for which the intersection $J_+^\Omega(p) \cap J_-^\Omega(q)$ is not compact, see
Fig. 2.22.

Example 15. The n-dimensional anti-deSitter spacetime $(\mathbb{R} \times S_+^{n-1}, g_{AdS})$ is not
globally hyperbolic (Fig. 2.23). As seen in Example 5, a curve in $M = \mathbb{R} \times S_+^{n-1}$
is causal with respect to g_{AdS} if and only if it is so with respect to the Lorentzian
metric $-dt^2 + \mathrm{can}_{S_+^{n-1}}$. Hence for both g_{AdS} and $-dt^2 + \mathrm{can}_{S_+^{n-1}}$. one gets the same
causal futures and pasts. A similar picture as in Example 14 then shows that for
$p, q \in M$ the intersection $J_+^M(p) \cap J_-^M(q)$ need not be compact.

In general one does not know much about causal futures and pasts in spacetime.
For globally hyperbolic manifold one has the following lemma (see [1, Chap. 14,
Lemma 22]).

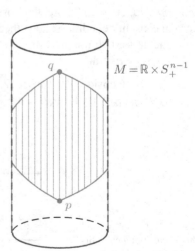

Fig. 2.23 $J_+^M(p) \cap J_-^M(q)$ is
not compact in anti-deSitter
spacetime

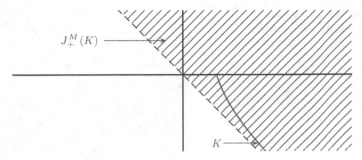

$J_+^M(K)$

K

Fig. 2.24 For K is closed $J_+^M(K)$ need not be open

Lemma 4. *In any globally hyperbolic manifold M the relation "$\leq$" is closed, i.e., whenever one has convergent sequences $p_i \to p$ and $q_i \to q$ in M with $p_i \leq q_i$ for all i, then one also has $p \leq q$.*

Therefore in globally hyperbolic manifolds for any $p \in M$ and any compact set $K \subset M$ one has that $J_\pm^M(p)$ and $J_\pm^M(K)$ are closed.

If K is only assumed to be closed, then $J_\pm^M(K)$ need not be closed. In Fig. 2.24 a curve K is shown which is closed as a subset and asymptotic to to a lightlike line in two-dimensional Minkowski space. Its causal future $J_+^M(K)$ is the open half-plane bounded by this lightlike line.

2.5 Cauchy Hypersurfaces

We recall that a piecewise C^1-curve in M is called *inextendible*, if no piecewise C^1-reparametrization of the curve can be continuously extended beyond any of the end points of the parameter interval.

Definition 10. *A subset S of a connected time-oriented Lorentzian manifold is called* achronal *(or* acausal*) if and only if each timelike (or causal, respectively) curve meets S at most once.*

A subset S of a connected time-oriented Lorentzian manifold is a Cauchy hypersurface *if each inextendible timelike curve in M meets S at exactly one point.*

Obviously every acausal subset is achronal, but the reverse is wrong. However, every achronal spacelike hypersurface is acausal (see [1, Chap. 14, Lemma 42]). Any Cauchy hypersurface is achronal. Moreover, it is a closed topological hypersurface and it is hit by each inextendible causal curve in at least one point. Any two Cauchy hypersurfaces in M are homeomorphic. Furthermore, the causal future and past of a Cauchy hypersurface is past- and future-compact, respectively. This is a consequence of, e.g., [1, Chap. 14, Lemma 40].

Example 16. In Minkowski space $(\mathbb{R}^n, g_{Mink})$ consider a spacelike hyperplane A_1, hyperbolic spaces $A_2 = \{x = (x_1, \ldots, x_n) \mid \langle\!\langle x, x \rangle\!\rangle = -1$ and $x_1 > 0\}$ and $A_3 = \{x = (x_1, \ldots, x_n) \mid \langle\!\langle x, x \rangle\!\rangle = 0, x_1 \geq 0\}$. Then all A_1, A_2, and A_3 are achronal, but only A_1 is a Cauchy hypersurface; see Fig. 2.25.

Fig. 2.25 Achronal subsets
A_1, A_2, and A_3 in Minkowski
space

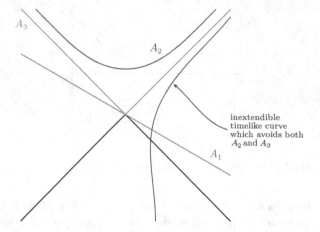

inextendible
timelike curve
which avoids both
A_2 and A_3

Example 17. Let (N, h) be a connected Riemannian manifold, let $I \subset \mathbb{R}$ be an open
interval and $f : I \to (0, \infty)$ a smooth function. Consider on $M = I \times N$ the
warped product metric $g = -dt^2 + f(t) \cdot h$. Then $\{t_0\} \times N$ is a Cauchy hypersurface
in (M, g) for any $t_0 \in I$ if and only if the Riemannian manifold (N, h) is complete
(compare [3, Lemma A.5.14]).

In particular, in any Robertson–Walker spacetime one can find a Cauchy hypersurface.

Example 18. Let N be exterior Schwarzschild spacetime N and B Schwarzschild
black hole, both of mass m, as defined in Example 4. Then for any $t_0 \in \mathbb{R}$ a Cauchy
hypersurface of N is given by $(2m, \infty) \times \{t_0\} \times S^2$, and in B one gets a Cauchy
hypersurface by $\{r_0\} \times \mathbb{R} \times S^2$ for any $0 < r_0 < 2m$.

Definition 11. *The* Cauchy development *(Fig. 2.26) of a subset S of a spacetime M
is the set $D(S)$ of points of M through which every inextendible causal curve in M
meets S, i.e.,*

$$D(S) = \{ p \in M \mid every\ inextendible\ causal\ curve\ passing\ through\ p\ meets\ S \}.$$

Remark 6. It follows from the definition that $D(D(S)) = D(S)$ for every subset
$S \subset M$. Hence if $T \subset D(S)$, then $D(T) \subset D(D(S)) = D(S)$.

Of course, if S is achronal, then every inextendible causal curve in M meets S at
most once. The Cauchy development $D(S)$ of every *acausal* hypersurface S is open,
see [1, Chap. 14, Lemma 43].

If $S \subset M$ is a Cauchy hypersurface, then obviously $D(S) = M$.

For a proof of the following proposition, see [1, Chap. 14, Thm. 38].

Proposition 3. *For any achronal subset $A \subset M$ the interior* int$(D(A))$ *of the
Cauchy development is globally hyperbolic (if nonempty).*

From this we conclude that a spacetime is globally hyperbolic if it possesses a
Cauchy hypersurface. In view of Examples 17 and 18, this shows that Robertson–

Fig. 2.26 Cauchy
development

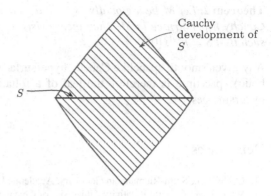

Walker spacetimes, Schwarzschild exterior spacetime, and Schwarzschild black hole are all globally hyperbolic.

The following theorem is very powerful and describes the structure of globally hyperbolic manifolds explicitly: they are foliated by *smooth spacelike* Cauchy hypersurfaces.

Theorem 1. *Let M be a connected time-oriented Lorentzian manifold. Then the following are equivalent:*

(1) M is globally hyperbolic.
(2) There exists a Cauchy hypersurface in M.
(3) M is isometric to $\mathbb{R} \times S$ with metric $-\beta dt^2 + g_t$ where β is a smooth positive function, g_t is a Riemannian metric on S depending smoothly on $t \in \mathbb{R}$ and each $\{t\} \times S$ is a smooth spacelike Cauchy hypersurface in M.

Proof. The crucial point in this theorem is that (1) implies (3). This has been shown by A. Bernal and M. Sánchez in [6, Theorem 1.1] using work of R. Geroch [7, Theorem 11]. See also [8, Preposition 6.6.8] and [2, p. 209] for earlier mentionings of this fact. That (3) implies (2) is trivial, and Proposition 3 provides the implication (2)⇒(1). □

Corollary 2. *On every globally hyperbolic Lorentzian manifold M there exists a smooth function $h : M \to \mathbb{R}$ whose gradient is past-directed timelike at every point and all of whose level sets are spacelike Cauchy hypersurfaces.*

Proof. Define h to be the composition $t \circ \Phi$ where $\Phi : M \to \mathbb{R} \times S$ is the isometry given in Theorem 1 and $t : \mathbb{R} \times S \to \mathbb{R}$ is the projection onto the first factor. □

Such a function h on a globally hyperbolic Lorentzian manifold is called a *Cauchy time function*. Note that a Cauchy time function is strictly monotonically increasing along any future-directed causal curve.

We conclude with an enhanced form of Theorem 1, due to A. Bernal and M. Sánchez (see [9, Theorem 1.2]).

Theorem 2. *Let M be a globally hyperbolic manifold and S be a spacelike smooth Cauchy hypersurface in M. Then there exists a Cauchy time function $h : M \to \mathbb{R}$ such that $S = h^{-1}(\{0\})$.* □

Any given smooth spacelike Cauchy hypersurface in a (necessarily globally hyperbolic) spacetime is therefore the leaf of a foliation by smooth spacelike Cauchy hypersurfaces.

References

1. O'Neill, B.: Semi-Riemannian Geometry. Academic Press, San Diego (1983)
2. Wald, R.M.: General Relativity. University of Chicago Press, Chicago (1984)
3. Bär, C., Ginoux, N., Pfäffle, F.: Wave equations on Lorentzian manifolds and quantization. EMS Publishing House, Zürich (2007)
4. Friedlander, F.: The wave equation on a curved space-time. Cambridge University Press, Cambridge (1975)
5. Leray, J.: Hyperbolic Differential Equations. Mimeographed Lecture Notes, Princeton (1953)
6. Bernal, A.N., Sánchez, M.: Smoothness of time functions and the metric splitting of globally hyperbolic spacetimes. Commun. Math. Phys. **257**, 43 (2005)
7. Geroch, R.: Domain of dependence. J. Math. Phys. **11**, 437 (1970)
8. Ellis, G.F.R., Hawking, S.W.: The large scale structure of space-time. Cambridge University Press, London-New York (1973)
9. Bernal, A.N., Sánchez, M.: Further results on the smoothability of Cauchy hypersurfaces and Cauchy time functions. Lett. Math. Phys. **77**, 183 (2006)

Chapter 3
Linear Wave Equations

Nicolas Ginoux

3.1 Introduction

This chapter deals with linear wave equations on Lorentzian manifolds. We first
recall the physical origin of that equation which describes the propagation of a wave
in space. Consider $\mathbb{R}^3$ with its canonical cartesian coordinates and let $u(t, x)$ denote
the height of the wave at $x = (x_1, x_2, x_3) \in \mathbb{R}^3$ and at time $t \in \mathbb{R}$. Then u solves

$$\frac{\partial^2 u}{\partial t^2} = \frac{\partial^2 u}{\partial x_1^2} + \frac{\partial^2 u}{\partial x_2^2} + \frac{\partial^2 u}{\partial x_3^2},$$

i.e. $\Box u = 0$, where $\Box := \frac{\partial^2}{\partial t^2} - \sum_{j=1}^{3} \frac{\partial^2}{\partial x_j^2}$ is the so-called d'Alembert operator on
the four-dimensional Minkowski space $\mathbb{R}^4 = \mathbb{R} \times \mathbb{R}^3$.
What can be said about the solutions u to the wave equation or equivalently about
the kernel of $\Box$? The operator $\Box$ is obviously linear, so that its kernel is a vector
space. The functions $(t, x) \mapsto \cos(nt)\cos(nx_1)$, for n running over $\mathbb{Z}$, all belong to
$\mathrm{Ker}(\Box)$ so that it is infinite dimensional. However, if one prescribes the height and
the speed of the wave at some fixed time then it is well known (see also Sect. 3.5.3)
that the corresponding solution must be unique.

Our aim here is to handle wave equations associated with some kind of generalized d'Alembert operators on any Lorentzian manifold. In particular, we want to
discuss the local and global existence of solutions as well as give a short motivation
on how those provide the fundamental background for some quantization theory.

The first section makes the concept of generalized d'Alembert operator more
precise and recalls the central role of fundamental solutions for differential operators. Fundamental solutions for the d'Alembert operator on the Minkowski space
can be obtained from the so-called Riesz distributions: this is the object of Sect. 3.3.
They are called advanced or retarded fundamental solutions according to their support being contained in the causal future or past of the origin. Turning to "curved"

N. Ginoux (✉)
NWF I – Mathematik, Universität Regensburg, D-93040 Regensburg, Germany
e-mail: Nicolas.Ginoux@mathematik.uni-regensburg.de

Ginoux, N.: *Linear Wave Equations*. Lect. Notes Phys. **786**, 59–84 (2009)
DOI 10.1007/978-3-642-02780-2_3   © Springer-Verlag Berlin Heidelberg 2009

spacetimes, i.e. to Lorentzian manifolds, there does not exist any analogue of Riesz distribution, at least globally. Nevertheless using normal coordinates it is always possible to transport Riesz distributions from the tangent space at a point to a neighbourhood of this point. The distributions obtained do not lead to local fundamental solutions for the classical d'Alembert operator in a straightforward manner (Sect. 3.4.1), however, combining linearly an infinite number of them, solving the wave equation formally (Sect. 3.4.2) and correcting the formal series using a cutoff function one obtains a local fundamental solution up to an error term (Proposition 2 in Sect. 3.4.3). General methods of functional analysis then allow one to get rid of this error term and construct true local fundamental solutions for any generalized d'Alembert operator (Corollary 2). Those fundamental solutions are in some sense near to the formal series from which they are constructed (Corollary 3).

The global aspect of the theory is based on a completely different approach. First it would be illusory to construct global fundamental solutions on any spacetime; therefore, we restrict the issue to globally hyperbolic spacetimes, which can be thought of as the analogue of complete Riemannian manifolds in the Lorentzian setting. In that case global fundamental solutions for generalized d'Alembert operators are provided by the solutions to the so-called Cauchy problem associated with such operators, see Sect. 3.5. After discussing uniqueness of fundamental solutions (Sect. 3.5.1) we show how local and then global solutions to the Cauchy problem can be constructed (Sects. 3.5.2 and 3.5.3) and global fundamental solutions be deduced from them (Sect. 3.5.4). Here it should be pointed out that the local existence of fundamental solutions (Sect. 3.4.3) actually enters this global construction in a crucial way since it provides (local) solutions to the so-called inhomogeneous wave equation; see Proposition 6.

We end this survey by introducing the concept of (advanced or retarded) Green's operators associated with generalized d'Alembert operators and by showing their elementary properties, in particular their tight relationship with fundamental solutions (Sect. 3.6). Green's operators constitute the starting point for the so-called local approach to quantization, since their existence together with a few additional assumptions on a given spacetime directly provide a C^*-algebra in a functorial way; see the last chapter for more details.

This chapter is intended as an introduction to the subject for students from the first or second university level. Only the main results and some ideas are presented; nevertheless, most proofs are left aside. We shall also exclusively deal with scalar operators, although all results of Sect. 3.4, 3.5 and 3.6 can be extended to generalized d'Alembert operators acting on sections of vector bundles. For a thorough and complete introduction to the topic as well as a list of references we refer to [1], on which this survey is widely based.

3.2 General Setting

In this section we describe the general frame in which we want to work.

3.2.1 Generalized Wave Equations

In the following and unless explicitly mentioned otherwise (M^n, g) will denote an n-dimensional Lorentzian manifold and $\mathbb{K} := \mathbb{R}$ or $\mathbb{C}$.

Definition 1. *A generalized d'Alembert operator on M is a linear differential operator of second-order P on M whose principal symbol is given by minus the metric.*

In the scalar setting, a generalized d'Alembert operator P is a linear differential operator of second order which can be written in local coordinates as follows:

$$P = -\sum_{i,j=0}^{n-1} g^{ij}(x) \frac{\partial^2}{\partial x_i \partial x_j} + \sum_{j=0}^{n-1} a_j(x) \frac{\partial}{\partial x_j} + b_1(x),$$

where a_j and b_1 are smooth $\mathbb{K}$-valued functions of x and $(g^{ij})_{i,j} := (g_{kl})_{k,l}^{-1}$. In particular, if P is a generalized d'Alembert operator on M then so is its formal adjoint P^*.

Examples

1. The d'Alembert operator of (M^n, g) is defined on smooth functions by

$$\Box f := -\mathrm{tr}_g(\mathrm{Hess}(f)),$$

where Hess $(f)(X, Y) := \langle \nabla_X \mathrm{grad} f, Y \rangle$ and tr_g is the trace w.r.t. the metric g. Here we denote as usual by ∇ the Levi-Civita covariant derivative on TM and by $\mathrm{grad} f$ the gradient vector field of the (real- or complex-valued) function f. In normal coordinates about $x \in M$

$$\Box f = -\mu_x^{-1} \sum_{j=0}^{n-1} \frac{\partial}{\partial x_j} (\mu_x (\mathrm{grad} f)_j),$$

with $\mu_x := |\det((g_{ij})_{i,j})|^{\frac{1}{2}}$; therefore, the principal symbol of $\Box$ is given by minus the metric. For instance on the Minkowski space $M = \mathbb{R}^n$ one has $\mu_x = 1$, hence

$$\Box f = -\sum_{j=0}^{n-1} \frac{\partial}{\partial x_j} \left(\varepsilon_j \frac{\partial f}{\partial x_j} \right) = \frac{\partial^2 f}{\partial x_0^2} - \sum_{j=1}^{n-1} \frac{\partial^2 f}{\partial x_j^2},$$

where $\varepsilon_0 := -1$ and $\varepsilon_j := 1$ for every $j \geq 1$.
2. The general form of a generalized d'Alembert operator is actually given by $\Box + a + b$, where a and b are linear differential operators of first and zero orders, respectively (b is a smooth function on M). In particular, the Klein–Gordon operator $\Box + m^2$, where $m > 0$ is a constant, is a generalized d'Alembert operator.

Other examples of generalized d'Alembert operators are given by the square of
any generalized Dirac operator on a Clifford bundle (see [2] for definitions) such
as the classical Dirac operator acting on spinors or the Euler operator acting on
differential forms. For the sake of simplicity we do not deal with vector bundles,
hence we restrict the whole discussion to scalar operators, i.e. to operators acting
on scalar-valued functions. From now on any differential operator will be implicitly
assumed to be scalar.

Definition 2. *Let P be a generalized d'Alembert operator on a Lorentzian manifold
M. The* wave equation *associated with P is*

$$Pu = f$$

for a given $f \in C^\infty(M, \mathbb{K})$.

We want to prove existence and uniqueness results – locally as well as globally –
for waves, i.e. for solutions $u \in C^\infty(M, \mathbb{K})$ to this generalized wave equation for
given data f lying in a particular class of functions. In this context we recall the
central role played by fundamental solutions.

3.2.2 Fundamental Solutions

We first recall what we need about distributions.

Definition 3. *The space of* $\mathbb{K}$*-valued distributions on M is defined as*

$$\mathcal{D}'(M, \mathbb{K}) := \{T : \mathcal{D}(M, \mathbb{K}) \longrightarrow \mathbb{K} \text{ linear and continuous}\},$$

where $\mathcal{D}(M, \mathbb{K}) := \{\varphi \in C^\infty(M, \mathbb{K}), \text{supp}(\varphi) \text{ compact}\}$ *denotes the space of*
$\mathbb{K}$*-valued test-functions on M.*

For the definition of the topology of $\mathcal{D}(M, \mathbb{K})$ we refer to Sect. 4.2.1 on page 86
and to [1, Chap. 1]. We next describe how functions can be understood as distributions and how differential operators act on distributions:

- For any fixed $f \in C^\infty(M, \mathbb{K})$ the map $\varphi \mapsto \int_M f(x)\varphi(x)dx$, $\mathcal{D}(M, \mathbb{K}) \to \mathbb{K}$
  defines a $\mathbb{K}$-valued distribution on M. Here and in the following we denote by
  dx the canonical measure associated with the metric g on M. We denote this
  distribution again by f, i.e. we identify $C^\infty(M, \mathbb{K})$ as a subspace of $\mathcal{D}'(M, \mathbb{K})$.
- Given $T \in \mathcal{D}'(M, \mathbb{K})$ and a linear differential operator P on M one can define

$$PT[\varphi] := T[P^*\varphi]$$

for any $\varphi \in \mathcal{D}(M, \mathbb{K})$, where P^* denotes the formal adjoint of P. It is an easy
exercise using the definition of the topology of $\mathcal{D}(M, \mathbb{K})$ to show that $PT \in \mathcal{D}'(M, \mathbb{K})$.

Definition 4. *Let P be a generalized d'Alembert operator on M and $x \in M$. A fundamental solution for P at x on M is a distribution $F \in \mathcal{D}'(M, \mathbb{K})$ with*

$$PF = \delta_x,$$

where δ_x is the Dirac distribution in x (i.e. $\delta_x[\varphi] := \varphi(x)$ for all $\varphi \in \mathcal{D}(M, \mathbb{K})$).

What do fundamental solutions for P – which are distributions – have to do with solutions of the wave equation $Pu = f$ – which we wish to be smooth functions? The idea is that one can construct from the fundamental solutions for P solutions u to the wave equation $Pu = f$ for "any" given f. We state this in a bit more precise but purely formal manner, see, e.g. Proposition 6 for a situation where the following computation can be carried out under some further assumptions.

Assume namely that one had at each $x \in M$ a fundamental solution $F(x) \in \mathcal{D}'(M, \mathbb{K})$ for P at x on M and moreover that $F(x)$ depends continuously on x, meaning that $x \mapsto F(x)[\varphi]$ is a continuous function for all $\varphi \in \mathcal{D}(M, \mathbb{K})$. Fix $f \in C^\infty(M, \mathbb{K})$ and consider

$$u[\varphi] := \int_M f(x)F(x)[\varphi]dx$$

for all $\varphi \in \mathcal{D}(M, \mathbb{K})$. In other words, u is some kind of convolution product of f with F. Assume u to be a well-defined distribution, then for every $\varphi \in \mathcal{D}(M, \mathbb{K})$ one has

$$\begin{aligned} Pu[\varphi] &= u[P^*\varphi] \\ &= \int_M f(x)F(x)[P^*\varphi]dx \\ &= \int_M f(x)PF(x)[\varphi]dx \\ &= \int_M f(x)\varphi(x)dx \\ &= f[\varphi], \end{aligned}$$

that is, $Pu = f$ in the distributional sense. Thus every wave equation associated with P can be solved on M, at least in $\mathcal{D}'(M, \mathbb{K})$.

Therefore we momentarily forget about the wave equation itself and concentrate on the search for fundamental solutions. As we shall already see in the next section, if there exists one fundamental solution then there exist many of them in general, hence one has to fix an extra condition to single one particular fundamental solution out. The most natural condition here deals with the support of the fundamental solution (recall that the support of a distribution on a manifold M is the complementary subset of the largest open subset of M on which the distribution vanishes), namely assuming that M is a spacetime (connected time-oriented Lorentzian manifold), we look for fundamental solutions $F_+(x)$, $F_-(x) \in \mathcal{D}'(M, \mathbb{K})$ for P at x on M such that

$$\text{supp}(F_+(x)) \subset J_+^M(x) \qquad \text{resp.} \qquad \text{supp}(F_-(x)) \subset J_-^M(x), \qquad (3.1)$$

where $J_+^M(x)$ and $J_-^M(x)$ are the causal future and past of x in M, respectively. Such an $F_+(x)$ (resp. $F_-(x)$) will be called *advanced* (resp. *retarded*) fundamental solution for P at x on M. In physics this condition has to do with the finiteness of the propagation speed of a wave.

Remark. The most naive condition would be to require the support to be compact. There exists unfortunately no fundamental solution with compact support in general, as the example of $P = \square$ on M already shows. Indeed if such a distribution F existed it could be extended to a continuous linear form on $C^\infty(M, \mathbb{K})$ (see Chap. 4), hence any non-zero constant φ would satisfy

$$\varphi(x) = \square F[\varphi] = F[\underbrace{\square \varphi}_{0}] = 0,$$

which would be a contradiction. In particular there exists no fundamental solution for $\square$ on compact Lorentzian manifolds.

3.3 Riesz Distributions on the Minkowski Space

In this section we describe the fundamental solutions for the d'Alembert operator at 0 on the Minkowski space $(\mathbb{R}^n, \langle\!\langle \cdot, \cdot \rangle\!\rangle_0)$ (recall that $\langle\!\langle x, y \rangle\!\rangle_0 := -x_0 y_0 + \sum_{j=1}^{n-1} x_j y_j$ for all $x = (x_0, x_1, \ldots, x_{n-1})$ and $y = (y_0, y_1, \ldots, y_{n-1})$ in $\mathbb{R}^n$) for $n \geq 2$.

Definition 5. *For $\alpha \in \mathbb{C}$ with $\mathfrak{Re}(\alpha) > n$ let $R_+(\alpha)$ and $R_-(\alpha)$ be the functions defined on $\mathbb{R}^n$ by*

$$R_\pm(\alpha)(x) := \begin{cases} C(\alpha, n)\gamma(x)^{\frac{\alpha-n}{2}} & \text{if } x \in J_\pm(0) \\ 0 & \text{otherwise,} \end{cases}$$

where

$$\gamma := -\langle\!\langle \cdot, \cdot \rangle\!\rangle_0, \; C(\alpha, n) := \frac{2^{1-\alpha}\pi^{\frac{2-n}{2}}}{(\frac{\alpha}{2} - 1)!(\frac{\alpha-n}{2})!} \text{ and } z \mapsto (z - 1)!$$

is the Gamma function.

Recall that the Gamma function is defined by $\{z \in \mathbb{C}, \mathfrak{Re}(z) > 0\} \to \mathbb{C}, z \mapsto \int_0^\infty t^{z-1}e^{-t}dt$. It is a holomorphic nowhere vanishing function on $\{\mathfrak{Re}(\alpha) > 0\}$ and satisfies

$$z! = z \cdot (z - 1)! \qquad (3.2)$$

for every $z \in \mathbb{C}$ with $\mathfrak{Re}(z) > 0$.

3 Linear Wave Equations

The function $R_\pm(\alpha)$ is well defined because of $\gamma \geq 0$ on $J_\pm(0)$, it is continuous on $\mathbb{R}^n$ and C^k as soon as $\Re e(\alpha) > n + 2k$. For any fixed $\varphi \in \mathcal{D}(\mathbb{R}^n, \mathbb{C})$ the map $\alpha \mapsto R_\pm(\alpha)[\varphi]$ is holomorphic on $\{\Re e(\alpha) > n\}$. Moreover $R_\pm(\alpha)$ satisfies the first important property.

Lemma 1. *For all $\alpha \in \mathbb{C}$ with $\Re e(\alpha) > n$ one has*

$$\Box R_\pm(\alpha + 2) = R_\pm(\alpha). \tag{3.3}$$

In particular the map $\alpha \mapsto R_\pm(\alpha)$, $\{\Re e(\alpha) > n\} \to \mathcal{D}'(\mathbb{R}^n, \mathbb{C})$ can be holomorphically extended on $\mathbb{C}$ such that (3.3) holds for every $\alpha \in \mathbb{C}$.

Of course by holomorphic extension we mean that, for every fixed $\varphi \in \mathcal{D}(\mathbb{R}^n, \mathbb{C})$, the function $\alpha \mapsto R_\pm(\alpha)[\varphi]$ can be holomorphic extended on $\mathbb{C}$.

Proof. The identity (3.3) follows from the two following ones:

- 1st identity:

$$\gamma \cdot R_\pm(\alpha) = \alpha(\alpha - n + 2)R_\pm(\alpha + 2). \tag{3.4}$$

Proof of (3.4): Both l.h.s and r.h.s vanish outside $J_\pm(0)$ so that we just have to prove the identity on $J_\pm(0)$. By Definition 6 one has on $J_\pm(0)$:

$$\gamma \cdot R_\pm(\alpha) = C(\alpha, n)\gamma^{\frac{\alpha+2-n}{2}}$$
$$= \frac{C(\alpha, n)}{C(\alpha + 2, n)}R_\pm(\alpha + 2),$$

with

$$\frac{C(\alpha, n)}{C(\alpha + 2, n)} = \frac{2^{1-\alpha}\pi^{\frac{2-n}{2}}}{(\frac{\alpha}{2} - 1)!(\frac{\alpha-n}{2})!} \cdot \frac{(\frac{\alpha}{2} - 1 + 1)!(\frac{\alpha+2-n}{2})!}{2^{1-\alpha-2}\pi^{\frac{2-n}{2}}}$$

$$\overset{(3.2)}{=} 4 \cdot \frac{\alpha}{2} \cdot \frac{\alpha + 2 - n}{2}$$

$$= \alpha(\alpha - n + 2), \tag{3.5}$$

which concludes the proof of (3.4).

- 2nd identity: for every $X \in \mathbb{R}^n$,

$$\partial_X\gamma \cdot R_\pm(\alpha) = 2\alpha \, \partial_X R_\pm(\alpha + 2). \tag{3.6}$$

Proof of (3.6): From its definition, $R_\pm(\alpha + 2)$ is C^1 on $\mathbb{R}^n$. We show the identity on $I_\pm(0)$. On this domain and for every $X \in \mathbb{R}^n$,

$$\partial_X\gamma \cdot R_\pm(\alpha) = C(\alpha, n)\gamma^{\frac{\alpha-n}{2}} \cdot \partial_X\gamma$$

$$= \frac{2C(\alpha, n)}{\alpha + 2 - n}\partial_X(\gamma^{\frac{\alpha+2-n}{2}}),$$

with

$$\frac{2C(\alpha, n)}{\alpha + 2 - n} \overset{(3.5)}{=} 2\alpha C(\alpha + 2, n),$$

so that

$$\partial_X \gamma \cdot R_\pm(\alpha) = 2\alpha C(\alpha + 2, n)\partial_X(\gamma^{\frac{\alpha+2-n}{2}})$$
$$= 2\alpha \, \partial_X R_\pm(\alpha + 2),$$

which is (3.6).

We now prove (3.3). Let first $\alpha \in \mathbb{C}$ with $\Re e(\alpha) > n + 2$. The function $R_\pm(\alpha + 2)$ is then C^2 on $\mathbb{R}^n$ and for every $X \in \mathbb{R}^n$,

$$\partial_X^2 R_\pm(\alpha + 2) \overset{(3.6)}{=} \frac{1}{2\alpha}\partial_X(\partial_X \gamma \cdot R_\pm(\alpha))$$
$$= \frac{1}{2\alpha}\left(\partial_X^2 \gamma \cdot R_\pm(\alpha) + \partial_X \gamma \cdot \partial_X R_\pm(\alpha)\right)$$
$$\overset{(3.6)}{=} \frac{1}{2\alpha}\left(\partial_X^2 \gamma \cdot R_\pm(\alpha) + \frac{1}{2(\alpha - 2)}(\partial_X \gamma)^2 \cdot R_\pm(\alpha - 2)\right),$$

with $(\partial_X \gamma)_x = -2\langle\!\langle x, X\rangle\!\rangle_0$ and $\partial_X^2 \gamma = 2\gamma(X)$ for any $x \in \mathbb{R}^n$. Hence, if $(e_j)_{0 \le j \le n}$ denotes the canonical basis of $\mathbb{R}^n$,

$$\Box R_\pm(\alpha + 2) = -\sum_{j=0}^{n} \varepsilon_j \frac{\partial^2 R_\pm(\alpha + 2)}{\partial x_j^2}$$
$$= -\frac{1}{2\alpha}\sum_{j=0}^{n}\left(2\varepsilon_j \gamma(e_j)R_\pm(\alpha) + \frac{2}{\alpha - 2}\varepsilon_j\langle\!\langle \cdot, e_j\rangle\!\rangle_0^2 R_\pm(\alpha - 2)\right)$$
$$= -\frac{1}{\alpha}\left(-nR_\pm(\alpha) - \frac{1}{\alpha - 2}\gamma \cdot R_\pm(\alpha - 2)\right)$$
$$\overset{(3.4)}{=} \frac{1}{\alpha}\left(nR_\pm(\alpha) + (\alpha - n)R_\pm(\alpha)\right)$$
$$= R_\pm(\alpha),$$

where, as usual, $\varepsilon_0 = \langle\!\langle e_0, e_0\rangle\!\rangle_0 = -1$ and $\varepsilon_j = \langle\!\langle e_j, e_j\rangle\!\rangle_0 = 1$ for every $1 \le j \le n - 1$. This proves (3.3) for $\Re e(\alpha) > n + 2$. It follows from the holomorphic dependence in α of both distributions $R_\pm(\alpha)$ and $\Box R_\pm(\alpha+2)$ that (3.3) must actually hold on the whole domain $\{\Re e(\alpha) > n\}$.

Equation (3.3) allows one to define inductively $R_\pm(\alpha)$ for every α with $\Re e(\alpha) > n - 2k$ with $k \in \mathbb{N}$. Indeed one can define the distribution $R_\pm(\alpha) := \Box R_\pm(\alpha + 2)$ for all $\alpha \in \mathbb{C}$ with $\Re e(\alpha) > n - 2$. For $\Re e(\alpha) > n$ this is of course not a definition but simply coincides with (3.3). Fix now $\varphi \in \mathcal{D}(\mathbb{R}^n, \mathbb{C})$. Since $\alpha \mapsto R_\pm(\alpha)[\varphi]$ is holomorphic on $\{\Re e(\alpha) > n\}$ then so is $\alpha \mapsto \Box R_\pm(\alpha)[\varphi]$. Hence the extension of $\alpha \mapsto R_\pm(\alpha)[\varphi]$ onto $\{\Re e(\alpha) > n - 2\}$ is again holomorphic and (3.3) is again trivially satisfied for those α. This shows the first step of the induction and achieves the proof of Lemma 1. $\Box$

Definition 6. *The* advanced *(resp. the* retarded*) Riesz distribution to the parameter* $\alpha \in \mathbb{C}$ *is defined to be* $R_+(\alpha)$ *(resp.* $R_-(\alpha)$*).*

The second important properties for our purpose are the following.

Lemma 2. *The Riesz distributions satisfy*

1. for any $\alpha \in \mathbb{C}$ *one has* $\mathrm{supp}(R_\pm(\alpha)) \subset J_\pm(0)$.
2. $R_\pm(0) = \delta_0$, *the Dirac distribution at the origin.*

The first assertion follows directly from the definition of the Riesz distributions and from $\Box$ being a differential operator. The second one requires a more technical and detailed study of the distribution $R_\pm(2)$, we refer to [1, Proposition 1.2.4] for a proof. Note also that, although $R_\pm(\alpha)$ is complex valued on $\mathcal{D}(M, \mathbb{C})$, its restriction to $\mathcal{D}(M, \mathbb{R})$ for real α gives a real-valued distribution.

As a consequence of Lemmas 1 and 2 we obtain the following.

Corollary 1. *The Riesz distribution* $R_\pm(2)$ *satisfies*

$$
\left|
\begin{array}{l}
\Box R_\pm(2) \quad = \delta_0 \\[2ex]
\mathrm{supp}(R_\pm(2)) \subset J_\pm(0).
\end{array}
\right.
$$

In particular $R_+(2)$ *(resp.* $R_-(2)$*) is an advanced (resp. retarded) fundamental solution for* $\Box$ *at* 0 *on* $\mathbb{R}^n$.

Remark. The set of fundamental solutions for a generalized d'Alembert operator P at a point is an affine subspace of $\mathcal{D}'(M, \mathbb{K})$ with direction $\mathrm{Ker}(P)$. For $M = \mathbb{R}^n$, since $\mathrm{Ker}(\Box)$ contains all constant extensions of harmonic functions on the spacelike slice $\mathbb{R}^{n-1}$, the space of fundamental solutions for $\Box$ at 0 on $\mathbb{R}^n$ is at least two dimensional for $n = 2$ and is infinite dimensional for $n \geq 3$ (remember that holomorphic functions on $\mathbb{C}$ are harmonic). This shows evidence that there exist significantly more than one fundamental solution as soon as there is one. Actually even if one keeps the support conditions (3.1) there may exist more than one fundamental solution. Consider, for instance, $P := \Box$ on $M := \mathbb{R} \times] - 1, 1[\subset \mathbb{R}^2$ with the induced Lorentzian metric. The restriction F_+ of $R_+(2)$ on M is an advanced fundamental solution for $\Box$ at 0 on M. On the other hand the map $\varphi \mapsto R_+(2)[\varphi \circ ((0, 3) + \mathrm{Id}]$, $\mathcal{D}(\mathbb{R}^2, \mathbb{C}) \to \mathbb{C}$ defines a distribution on $\mathbb{R}^2$ which is obviously an advanced fundamental solution for $\Box$ at $x = (0, 3)$ on $\mathbb{R}^2$, thus its restriction G_+ onto M lies in $\mathrm{Ker}(\Box)$ with support contained in $J_+(3) \cap M$ which is again a subset of $J_+^M(0)$. Therefore $F_+ + G_+$ is another advanced fundamental solution for $\Box$ at 0 in M.

However we shall see in Corollary 4 that, on $\mathbb{R}^n$, there exists exactly one advanced and one retarded fundamental solution for $\Box$ at 0.

3.4 Local Fundamental Solutions

In this section we come back to the general setting and construct local fundamental solutions for any generalized d'Alembert operator on any Lorentzian manifold.

3.4.1 Attempt

We first examine the case where $P = \square$ on M. From the local point of view the most naive attempt to obtain fundamental solutions for $\square$ on M consists in pulling the Riesz distributions $R_{\pm}(2)$ back from the tangent space at a point onto a neighbourhood of that point:

Definition 7. *Let Ω be a geodesically starshaped neighbourhood of a point x in a Lorentzian manifold (M^n, g). Let $\exp_x : \exp_x^{-1}(\Omega) \to \Omega$ be the exponential map and $\mu_x : \Omega \to \mathbb{R}$, $\mu_x := |\det((g_{ij})_{i,j})|^{\frac{1}{2}}$. The Riesz distribution at x on Ω to the parameter $\alpha \in \mathbb{C}$ is defined by*

$$R_{\pm}^{\Omega}(\alpha, x) : \mathcal{D}(\Omega, \mathbb{C}) \longrightarrow \mathbb{C}$$
$$\varphi \longmapsto R_{\pm}(\alpha)[(\mu_x \varphi) \circ \exp_x],$$

where $R_{\pm}(\alpha)$ denotes the Riesz distribution to the parameter α.

The factor μ_x enters the definition of $R_{\pm}^{\Omega}(\alpha, x)$ in order to take the difference between the volume form of M and that of $T_x M$ (w.r.t. g_x) into account: indeed $\mathrm{dvol}_g = \mu_x (\exp_x^{-1})^* \mathrm{dvol}_{g_x}$ on Ω.

By definition $R_{\pm}^{\Omega}(\alpha, x)$ is a distribution on Ω. It can be relatively easily deduced from its definition and from Lemmas 1 and 2 that $R_{\pm}^{\Omega}(\alpha, x)$ satisfies the following [1, Proposition 1.4.2]:

Lemma 3. *Let Ω be a geodesically starshaped neighbourhood of a point x in a Lorentzian manifold (M^n, g). Then the Riesz distributions at x on Ω satisfy*

1. $R_{\pm}^{\Omega}(0, x) = \delta_x$.
2. $\mathrm{supp}(R_{\pm}^{\Omega}(\alpha, x)) \subset J_{\pm}^{\Omega}(x)$.
3. $\square R_{\pm}^{\Omega}(\alpha + 2, x) = (\frac{\square \Gamma_x - 2n}{2\alpha} + 1) R_{\pm}^{\Omega}(\alpha, x)$ *for every* $\alpha \neq 0$, *where* $\Gamma_x := \gamma \circ \exp_x^{-1}$ *on* Ω.

Although the first two properties make $R_{\pm}^{\Omega}(2, x)$ a good candidate to become a fundamental solution for $\square$ at x, the third one sweeps this hope away since the term $\square \Gamma_x - 2n$ does not vanish in general. Therefore one has to look for another approach to find fundamental solutions for $\square$.

3.4.2 Formal Ansatz

Considering again any generalized d'Alembert operator P we fix a point $x \in M$ and a geodesically starshaped neighbourhood of x in (M^n, g). We look for fundamental solutions for P of the form

$$T_\pm := \sum_{k=0}^{\infty} V_x^k \cdot R_\pm^\Omega (2k + 2, x),$$

where for each k, V_x^k is a smooth coefficient depending on x. Of course this series is a priori only formal. Nevertheless if one plugs it into the equation $PT_\pm = \delta_x$, differentiates it termwise, uses relations satisfied by the Riesz distributions such as (3.4), (3.6) or Lemma 3 and identifies the coefficients standing in front of the $R_\pm^\Omega (2k + 2, x)$ then one obtains [1, Sect. 2.1]

$$\nabla_{\mathrm{grad}\,\Gamma_x} V_x^k - (\frac{1}{2}\Box\Gamma_x - n + 2k)V_x^k = 2kP V_x^{k-1} \tag{3.7}$$

for every $k \geq 1$ as well as $V_{x,x}^0 = 1$. This leads to the following.

Definition 8. *Let $\Omega \subset M$ be convex. A sequence of Hadamard coefficients for P on Ω is a sequence $(V^k)_{k \geq 0}$ of $C^\infty(\Omega \times \Omega, \mathbb{C})$ which fulfills (3.7) and $V_{x,x}^0 = 1$, for all $x \in \Omega$ and $k \geq 1$, where we denote by $V_x^k := V_{x,\cdot}^k \in C^\infty(\Omega, \mathbb{C})$.*

The Equation (3.7) that Hadamard coefficients must satisfy turns out to be a singular differential equation and can be solved without any further assumption [1, Sects. 2.2 and 2.3]. For the sake of simplicity we give a formula for the Hadamard coefficients only in the case where the operator P has no term of first order (the general formula involves the parallel transport of the connection which is canonically associated with P, see [1, Lemmas 1.5.5 and 2.2.2]).

Proposition 1. *Let $\Omega \subset M$ be a convex open subset in a Lorentzian manifold (M^n, g) and P be a generalized d'Alembert operator on M of the form $P = \Box + b$, where $b \in C^\infty(M, \mathbb{K})$. Then there exists a unique sequence of Hadamard coefficients for P on Ω. It is given for all $x, y \in \Omega$ by*

$$V_{x,y}^0 = \mu_x^{-\frac{1}{2}}(y)$$

and, for all $k \geq 1$,

$$V_{x,y}^k = -k\mu_x^{-\frac{1}{2}}(y)\int_0^1 \mu_x^{\frac{1}{2}}(\Phi(y,s))s^{k-1} \cdot (P_{(2)}V_x^{k-1}(\Phi(y,s)))ds,$$

where $\Phi(y,s) := \exp_x(s\exp_x^{-1}(y))$, $\Phi : \Omega \times [0, 1] \to \Omega$.

The index "(2)" in $P_{(2)}V_x^{k-1}$ stands for P acting on $z \mapsto V^{k-1}(x, z)$. The existence of Hadamard coefficients leads to the following definition.

Definition 9. *Let $\Omega \subset M$ be a convex open subset in a Lorentzian manifold (M^n, g) and P be a generalized d'Alembert operator on M. Let $(V^k)_{k\geq 0}$ be the sequence of Hadamard coefficients for P on Ω. The (advanced or retarded) formal fundamental solution for P at $x \in \Omega$ is the formal series*

$$R_{\pm}^{\Omega}(x) := \sum_{k=0}^{\infty} V_x^k \cdot R_{\pm}^{\Omega}(2k + 2, x).$$

3.4.3 Exact Local Fundamental Solutions

The existence of Hadamard coefficients still does not provide any (local) fundamental solution, since the series defining $R_{\pm}^{\Omega}(x)$ may diverge. The idea presented here for the construction of local fundamental solutions (which is that of [1]) consists in keeping the first terms of the formal fundamental solutions unchanged while multiplying the higher ones by a cutoff function.

More precisely, consider again a convex open subset Ω' in M. Let $\sigma : \mathbb{R} \to [0, 1]$ be a smooth function with $\mathrm{supp}(\sigma) \subset [-1, 1]$ and $\sigma_{|[-\frac{1}{2}, \frac{1}{2}]} = 1$. Fix an integer $N \geq \frac{n}{2}$ (this is just to ensure $R_{\pm}^{\Omega}(2k + 2, x)$ be a continuous function for any $k \geq N$) and a sequence $(\varepsilon_j)_{j \geq N}$ of positive real numbers. Set

$$\widetilde{R}_{\pm}(x) := \sum_{j=0}^{N-1} V_x^j \cdot R_{\pm}^{\Omega'}(2j + 2, x) + \sum_{j=N}^{\infty} \sigma\left(\frac{\Gamma_x}{\varepsilon_j}\right) \cdot V_x^j \cdot R_{\pm}^{\Omega'}(2j + 2, x) \quad (3.8)$$

for every $x \in \Omega$ (recall that $\Gamma_x := \gamma \circ \exp_x^{-1}$ with $\gamma := -\langle\!\langle \cdot, \cdot \rangle\!\rangle_0$). The identity (3.8) does not a priori define a fundamental solution since it does not even define a distribution. However, for ε_j small enough both conditions are almost fulfilled (see [1, Lemmas 2.4.2–2.4.4]):

Proposition 2. *Let $\Omega' \subset M$ be convex and $\Omega \subset\subset \Omega'$ be relatively compact. Fix an integer $N \geq \frac{n}{2}$. Then there exists a sequence $(\varepsilon_j)_{j \geq N}$ of positive real numbers such that, for all $x \in \overline{\Omega}$, $\widetilde{R}_{\pm}(x)$ defines a distribution on Ω satisfying*

(a) $P_{(2)}\widetilde{R}_{\pm}(x) - \delta_x = K_{\pm}(x, \cdot)$, *where* $K_{\pm} \in C^{\infty}(\overline{\Omega} \times \overline{\Omega}, \mathbb{C})$,
(b) $\mathrm{supp}(\widetilde{R}_{\pm}(x)) \subset J_{\pm}^{\Omega'}(x)$,
(c) $y \mapsto \widetilde{R}_{\pm}(y)[\varphi]$ *is smooth on Ω for all* $\varphi \in \mathcal{D}(\Omega, \mathbb{C})$.

In other words, choosing suitably the ε_j leads to a distribution depending smoothly on the base point and which is near to a fundamental solution in the sense that the difference $P_{(2)}\widetilde{R}_{\pm}(x) - \delta_x$ is a smooth function. How to obtain now a "true" fundamental solution? The main idea is to get rid of the error term using methods of functional analysis, namely setting, for all $u \in C^0(\overline{\Omega}, \mathbb{C})$,

$$\mathcal{K}_\pm u := \int_\Omega K_\pm(\cdot, y)u(y)dy,$$

the identity (a) of Proposition 2 can be rewritten in the form

$$P_{(2)}\widetilde{R}_\pm(\cdot)[\varphi] = (\mathrm{Id} + \mathcal{K}_\pm)\varphi$$

for all $\varphi \in \mathcal{D}(\Omega, \mathbb{C})$. One can therefore look for an *inverse* to the operator $\mathrm{Id} + \mathcal{K}_\pm$. It is well known that, given a bounded endomorphism A of a Banach space, the operator $\mathrm{Id} + A$ is invertible as soon as $\|A\| < 1$. This is the main idea underlying the following proposition (see [1, Lemma 2.4.8]).

Proposition 3. *Let* $\Omega \subset\subset \Omega'$ *be a relatively compact causal domain in* Ω' *and assume that* $\mathrm{Vol}(\overline{\Omega}) \cdot \|K_\pm\|_{C^0(\overline{\Omega}\times\overline{\Omega})} < 1$. *Then* $\mathrm{Id} + \mathcal{K}_\pm$ *is an isomorphism* $C^k(\overline{\Omega}, \mathbb{C}) \to C^k(\overline{\Omega}, \mathbb{C})$ *for all* $k \in \mathbb{N}$.

Setting

$$F_\pm^\Omega(\cdot)[\varphi] := (\mathrm{Id} + \mathcal{K}_\pm)^{-1}(y \mapsto \widetilde{R}_\pm(y)[\varphi])$$

for all $\varphi \in \mathcal{D}(\Omega, \mathbb{C})$, we really obtain what we wanted: for any $x \in \Omega$ and $\varphi \in \mathcal{D}(\Omega, \mathbb{C})$,

$$\begin{aligned}
(PF_\pm^\Omega(x))[\varphi] &= F_\pm^\Omega(x)[P^*\varphi] \\
&= \{(\mathrm{Id} + \mathcal{K}_+)^{-1}(y \mapsto \widetilde{R}_\pm(y)[P^*\varphi])\}(x) \\
&= \{(\mathrm{Id} + \mathcal{K}_\pm)^{-1}(\underbrace{y \mapsto P_{(2)}\widetilde{R}_\pm(y)[\varphi])}_{(\mathrm{Id}+\mathcal{K}_\pm)\varphi}\}(x) \\
&= \varphi(x),
\end{aligned}$$

that is, $PF_\pm^\Omega(x) = \delta_x$. The other properties $F_\pm^\Omega(x)$ should satisfy can be relatively easily checked, hence we can state the following.

Proposition 4. *Under the assumptions of* Proposition 3 *the map* $\varphi \mapsto F_+^\Omega(x)[\varphi]$ *is an advanced fundamental solution on* Ω *for* P *at* $x \in \Omega$ *and the map* $\varphi \mapsto F_-^\Omega(x)[\varphi]$ *is a retarded one.*

To sum up:

Corollary 2. *Let* P *be a generalized d'Alembert operator on a Lorentzian manifold* (M^n, g). *Then every point of* M *possesses a relatively compact causal neighbourhood* Ω *such that, for every* $x \in \Omega$, *there exist fundamental solutions* $F_\pm^\Omega(x)$ *on* Ω *for* P *at* x *satisfying*

(a) $\mathrm{supp}(F_\pm^\Omega(x)) \subset J_\pm^\Omega(x)$ *and*
(b) $x \mapsto F_\pm^\Omega(x)[\varphi]$ *is smooth for all* $\varphi \in \mathcal{D}(\Omega, \mathbb{C})$.

3.4.4 Comparison of Formal and Exact Local Fundamental Solutions

In this section we show that the formal fundamental solutions obtained in Sect. 3.4.2 is asymptotic to the true one along the light cone. More precisely [1, Proposition 2.5.1]:

Proposition 5. *Let* $\Omega \subset\subset \Omega'$ *be as in* Proposition 3. *Fix* $N \geq n/2$ *and for* $k \in \mathbb{N}$ *set*

$$\mathcal{R}_{\pm}^{N+k}(x) := \sum_{j=0}^{N+k-1} V_x^j \cdot R_{\pm}^{\Omega'}(2j+2, x),$$

where $(V^j)_{j \geq 0}$ *is the sequence of Hadamard coefficients for* P *on* Ω'. *Then for every* $k \in \mathbb{N}$ *the map*

$$(x, y) \mapsto (F_{\pm}^{\Omega}(x) - \mathcal{R}_{\pm}^{N+k}(x))(y)$$

is a C^k-*function on* $\Omega \times \Omega$, *where* $F_{\pm}^{\Omega}$ *is given by* Proposition 4.

This a strong statement, since both $\mathcal{R}_{\pm}^{N+k}(x)$ and $F_{\pm}^{\Omega}(x)$ are singular along the light cone $\{y \in \Omega \mid \Gamma_x(y) = 0\}$ based at x (see [1, Proposition 1.4.2]). Using an elementary argument of differential geometry [1, Lemma 2.5.4] one can deduce the following.

Corollary 3. *Under the assumptions of* Proposition 5, *there exists for every* $k \in \mathbb{N}$ *a constant* C_k *such that*

$$\|(F_{\pm}^{\Omega}(x) - \mathcal{R}_{\pm}^{N+k}(x))(y)\| \leq C_k \cdot |\Gamma_x(y)|^k$$

for all $(x, y) \in \Omega \times \Omega$.

3.5 The Cauchy Problem and Global Fundamental Solutions

In this section we want to construct global fundamental solutions. A naive idea would consist in taking the local fundamental solutions constructed above and gluing them together using a partition of unity. A quick reflection convinces one of the difficulties which then may occur; namely it is already not clear which equation should be solved in each coordinate patch not containing the point at which the fundamental solutions are sought after. Studying this question in more detail one immediately observes that the global topology and geometry of the manifold could set up serious problems. As we have already seen at the end of Sect. 3.2.2 there cannot exist any fundamental solution for $\square$ on compact spacetimes. Even if the manifold is not compact the possible existence of closed or almost-closed causal curves can make the very definition of fundamental solutions ill-posed; indeed it could theoretically happen that a wave overlaps itself after finite time.

Since we want to avoid this kind of situation we have to first properly choose our Lorentzian manifolds. There exists a "good" class of Lorentzian manifolds in this respect, which are called *globally hyperbolic* (see Definition 9 on page 53). We restrict the discussion of the issue to the globally hyperbolic setting, although uniqueness as well as existence results may each be extended to broader classes of spacetimes, see [1, Sects 3.1 and 3.5].

In this case we make what at first seems to be a detour: we solve the so-called *Cauchy problem*, which in analogy with ordinary differential equations consists in solving a wave equation fixing initial conditions on a subset of the manifold. Since generalized d'Alembert operators are differential operators of second order, two conditions have to be fixed:

Definition 10. *Let P be a generalized d'Alembert operator on a globally hyperbolic spacetime (M^n, g) and $S \subset M$ be a (smooth) spacelike hypersurface with unit normal vector field v. Let $f \in C^\infty(M, \mathbb{K})$ and $u_0, u_1 \in C^\infty(S, \mathbb{K})$. The* Cauchy problem *for P with Cauchy data (f, u_0, u_1) is the system of equations*

$$\left| \begin{array}{l} Pu = f \quad on \ M \\ u_{|S} = u_0 \\ \partial_v u = u_1 \quad on \ S. \end{array} \right.$$

We shall be interested in solving the Cauchy problem in $C^\infty(M, \mathbb{K})$ and with compactly supported data (see [3] for less regular solutions). The link with fundamental solutions will be explained in Sect. 3.5.4.

3.5.1 Uniqueness of Fundamental Solutions

We first show the uniqueness of advanced and retarded fundamental solutions at a point on globally hyperbolic spacetimes. One of the main ingredients involved is the local solvability of the following inhomogeneous wave equation.

Proposition 6. *Under the assumptions of* Proposition 3, *there exists for every* $v \in \mathcal{D}(\Omega, \mathbb{C})$ *a function* $u_\pm \in C^\infty(\Omega, \mathbb{C})$ *such that*

$$\left| \begin{array}{l} Pu_\pm = v \\ \mathrm{supp}(u_\pm) \subset J^\Omega_\pm(\mathrm{supp}(v)). \end{array} \right.$$

Sketch of proof. It follows from Proposition 4 that, for every $x \in \Omega$, there exist fundamental solutions $F^\Omega_\pm(x)$ for P at x on Ω. As in Sect. 3.2.2 we set

$$u_{\pm}[\varphi] := \int_{\Omega} v(x) F_{\pm}^{\Omega}(x)[\varphi] dx$$

for every $\varphi \in \mathcal{D}(\Omega, \mathbb{C})$. There are three assertions to be shown.

- The map $u_{\pm} \colon \mathcal{D}(\Omega, \mathbb{C}) \to \mathbb{C}$ is a solution of $Pu_{\pm} = v$ in $\mathcal{D}'(\Omega, \mathbb{C})$: That $u_{\pm}$ defines a distribution follows from $F_{\pm}^{\Omega}(x)$ being one, from $x \mapsto F_{\pm}^{\Omega}(x)[\varphi]$ being smooth for every $\varphi \in \mathcal{D}(\Omega, \mathbb{C})$ and from an uniform estimate of the order of the distributions $F_{\pm}^{\Omega}(x)$ for x running in Ω, see [1, Lemma 2.4.4]. As for $Pu_{\pm} = v$ in the distributional sense, this is exactly the computation carried out in Sect. 3.2.2 and which is justified.
- The support condition: Let $\varphi \in \mathcal{D}(\Omega, \mathbb{C})$ be such that $u_{\pm}[\varphi] \neq 0$, then there exists an $x \in \Omega$ such that $F_{\pm}^{\Omega}(x)[\varphi]v(x) \neq 0$, which implies $\mathrm{supp}(\varphi) \cap \mathrm{supp}(F_{\pm}^{\Omega}(x)) \neq \varnothing$ and $x \in \mathrm{supp}(v)$. Hence $\mathrm{supp}(\varphi) \cap J_{\pm}^{\Omega}(x) \neq \varnothing$, i.e. $x \in J_{\mp}^{\Omega}(\mathrm{supp}(\varphi))$, so that $J_{\mp}^{\Omega}(\mathrm{supp}(\varphi)) \cap \mathrm{supp}(v) \neq \varnothing$, or equivalently $\mathrm{supp}(\varphi) \cap J_{\pm}^{\Omega}(\mathrm{supp}(v)) \neq \varnothing$, which was to be proved.
- The distribution $u_{\pm}$ is in fact a smooth section: This is the technical part of the proof, which actually relies not only on Corollary 2 but also on the explicit form of the local fundamental solutions on Ω, see [1, Sect. 2.6] for details.

$\square$

We state the main result of this section.

Theorem 1. *Let P be a generalized d'Alembert operator on a globally hyperbolic spacetime (M^n, g). Then every solution $u \in \mathcal{D}'(M, \mathbb{C})$ with past- or future-compact support of the equation $Pu = 0$ vanishes.*

Sketch of proof. Assume u has past-compact support (the other case is completely analogous). One has to show that $u[\varphi] = 0$ for every $\varphi \in \mathcal{D}(M, \mathbb{C})$. The idea is to apply Proposition 6 and solve the inhomogeneous wave equation to the operator P^* (which is of generalized d'Alembert type, see Sect. 3.2.1)

$$\left| \begin{array}{l} P^*\psi \quad = \varphi \\[2mm] \mathrm{supp}(\psi) \subset J_{-}^{\Omega}(\mathrm{supp}(\varphi)) \end{array} \right.$$

for any fixed φ with "small" support (small enough in order to Proposition 6 be applied). Since u has past-compact support one can hope for the intersection $\mathrm{supp}(u) \cap J_{-}^{\Omega}(\mathrm{supp}(\varphi))$ be compact and hence for

$$\begin{aligned} u[\varphi] &= u[P^*\psi] \\ &= \underbrace{Pu}_{0}[\psi] \\ &= 0, \end{aligned}$$

which would be the result. The whole work is to justify this computation as well as the assumptions on $\mathrm{supp}(\varphi)$, which requires global properties of the causality relation, see [1, Theorem 3.1.1]. This completes the sketch of proof of Theorem 1.   □

Since $J_+^M(K)$ (resp. $J_-^M(K)$) is past (resp. future) compact for any compact subset K of a globally hyperbolic spacetime M, we obtain the following.

Corollary 4. *Let P be a generalized d'Alembert operator on a globally hyperbolic spacetime (M^n, g) and $x \in M$. Then there exists at most one advanced (resp. retarded) fundamental solution for P at x.*

3.5.2 Cauchy Problem: Local Solvability

The existence of solutions to the Cauchy problem is a local-to-global construction. In this section we deal with the local aspect, which of course does not require global hyperbolicity of the manifold.

Theorem 2. *Let (M^n, g) be a spacetime and S be a smooth spacelike hypersurface with (timelike) unit normal vector field v. Then for each open subset Ω of M satisfying the hypotheses of Proposition 3 and such that $S \cap \Omega$ is a Cauchy hypersurface of Ω the following holds: for all $u_0, u_1 \in \mathcal{D}(S \cap \Omega, \mathbb{C})$ and each $f \in \mathcal{D}(\Omega, \mathbb{C})$, there exists a unique $u \in C^\infty(\Omega, \mathbb{C})$ with*

$$
\left|
\begin{aligned}
Pu &= f \\
u_{|s} &= u_0 \\
\partial_v u &= u_1.
\end{aligned}
\right.
$$

Furthermore $\operatorname{supp}(u) \subset J_+^\Omega(K) \cup J_-^\Omega(K)$, *where*

$$
K := \operatorname{supp}(u_0) \cup \operatorname{supp}(u_1) \cup \operatorname{supp}(f).
$$

Sketch of proof. Although Proposition 6 naturally enters the proof, it does not straightforwardly imply the result. One has to formulate a separate ansatz, namely writing Ω as the product $\mathbb{R} \times (S \cap \Omega)$ (which is possible since $S \cap \Omega$ is assumed to be a Cauchy hypersurface in Ω and because of Theorem 1), one looks for a solution of the form

$$
\sum_{j=0}^\infty t^j u_j(x),
$$

where $(t, x) \in \mathbb{R} \times (S \cap \Omega)$ and $u_j \in C^\infty(S \cap \Omega, \mathbb{C})$. As for the Hadamard coefficients one obtains inductive relations of the form $u_j = \mathcal{F}(u_0, u_1, \dots, u_{j-1})$ for every $j \geq 2$, which then stand as definition for the u_j in terms of u_0 and u_1 on the whole of $S \cap \Omega$. Coming back to the inhomogeneous equation, one introduces a cutoff function σ as in the construction of local fundamental solutions (see Sect. 3.4.3) and sets

$$\hat{u} := \sum_{j=0}^{\infty} \sigma \left(\frac{t}{\varepsilon_j} \right) t^j u_j,$$

with the $u_j \in C^{\infty}(S \cap \Omega, \mathbb{C})$ found above and a sequence $(\varepsilon_j)_{j \geq 0}$ of positive real numbers. Choosing the ε_j suitably small one obtains a smooth function $\hat{u}$ on Ω such that $P\hat{u} - f$ vanishes not on all Ω but at least on $S \cap \Omega$ at infinite order. Proposition 6 then provides smooth solutions $\tilde{u}_+$ and $\tilde{u}_-$ of the respective inhomogeneous problems on Ω:

$$\left|
\begin{array}{l}
P\tilde{u}_\pm = w_\pm \\[2mm]
\mathrm{supp}(\tilde{u}_\pm) \subset J_\pm^{\Omega}(\mathrm{supp}(w_\pm)),
\end{array}
\right.$$

where $w_\pm|_{J_\pm^{\Omega}(S \cap \Omega)} := P\hat{u} - f$ and vanishes on the rest of Ω. The last step consists in showing that $u_\pm := \hat{u} - \tilde{u}_\pm$ solves the wave equation $Pu_\pm = f$ on $J_\pm^{\Omega}(S \cap \Omega)$ and vanishes on $J_\mp^{\Omega}(S \cap \Omega)$. The function u defined by

$$u := \left|
\begin{array}{l}
u_+ \quad \text{on } J_+^{\Omega}(S \cap \Omega) \\[2mm]
u_- \quad \text{on } J_-^{\Omega}(S \cap \Omega)
\end{array}
\right.$$

is then a smooth function on Ω solving the requested Cauchy problem.

The uniqueness – which is actually needed for the last step of the existence just above – follows from an independent argument, which is an integral formula for solutions of the Cauchy problem on Ω, namely if $u \in C^{\infty}(\Omega, \mathbb{C})$ solves $Pu = 0$, then

$$\int_{\Omega} u(x) \varphi(x) dx = \int_{S \cap \Omega} (\partial_\nu (F^{\Omega}[\varphi]) u_0 - F^{\Omega}[\varphi] u_1) ds$$

for all $\varphi \in \mathcal{D}(\Omega, \mathbb{C})$, where $F^{\Omega}[\varphi] : \mathcal{D}(\Omega, \mathbb{C}) \to \mathbb{C}, \psi \mapsto \int_{\Omega} \varphi(x)(F_+^{\Omega}(x)[\psi] - F_-^{\Omega}(x)[\psi]) dx$ and ds is the induced measure on $S \cap \Omega$, see [1, Lemma 3.2.2]. Therefore if u_0 and u_1 vanish on $S \cap \Omega$ then u vanishes – as distribution and hence as function – on Ω.

The control of the support of the solution follows from the corresponding one for the inhomogeneous problem (Proposition 6) and from the integral formula just above. This completes the sketch of proof of Theorem 2. $\square$

Note that, since every point on a spacelike hypersurface S in M admits a basis of neighbourhoods Ω_j in M such that $S \cap \Omega_j$ is a Cauchy hypersurface of Ω_j (roughly speaking one just has to consider the Cauchy development of $S \cap U$ in an open subset U meeting S, see, e.g. [1, Lemma A.5.6]), Theorem 2 actually proves that the Cauchy problem with compactly supported data is always locally solvable.

3.5.3 Cauchy Problem: Global Solvability

We come to the central result of this survey.

Theorem 3. *Let P be a generalized d'Alembert operator on a globally hyperbolic spacetime M and $S \subset M$ be a spacelike Cauchy hypersurface in M with (timelike) unit normal ν.*

(i) For all $(f, u_0, u_1) \in \mathcal{D}(M, \mathbb{C}) \oplus \mathcal{D}(S, \mathbb{C}) \oplus \mathcal{D}(S, \mathbb{C})$, there exists a unique $u \in C^\infty(M, \mathbb{C})$ such that

$$\left|\begin{array}{l} Pu = f \\[2mm] u_{|_S} = u_0 \\[2mm] \partial_\nu u = u_1. \end{array}\right. \tag{3.9}$$

Moreover $\operatorname{supp}(u) \subset J_+^M(K) \cup J_-^M(K)$ *with*

$$K := \operatorname{supp}(u_0) \cup \operatorname{supp}(u_1) \cup \operatorname{supp}(f).$$

(ii) The map

$$\mathcal{D}(M, \mathbb{C}) \oplus \mathcal{D}(S, \mathbb{C}) \oplus \mathcal{D}(S, \mathbb{C}) \to C^\infty(M, \mathbb{C})$$
$$(f, u_0, u_1) \mapsto u,$$

where $u \in C^\infty(M, \mathbb{C})$ is the solution of (3.9), is linear continuous.

Sketch of proof. The existence of u in (i) , which is rather technical, is proved in two main steps. First one constructs a solution u in a strip $]-\varepsilon, \varepsilon[\times S$ (where M is identified with $\mathbb{R} \times S$) for some $\varepsilon > 0$: this is the easier step, since it roughly means gluing together local solutions obtained by Theorem 2 along the hypersurface $S \simeq \{0\} \times S$. There is only a finite number of them to be taken into account since $(J_+^M(K) \cup J_-^M(K)) \cap S$ is compact (outside this intersection u should vanish along S). In the second step one shows that u can be extended in the whole future and past of the strip. The core of the global theory lies here, namely using the local theory and the global hyperbolicity of M it can be shown that u can be continued in the future or past "independently" of the behaviour of the already existing u; in other words, no explosion can occur. We refer to [1, Theorem. 3.2.11] for a clean and thorough argumentation.

The uniqueness of u follows from another technical argument based on the local integral formula described in the proof of Theorem 2, see [1, Corollary. 3.2.4]. This shows (i).

Statement (ii), which should be interpreted as a stability result for waves (solutions of the Cauchy problem depend continuously on the data), is a not-so-direct application of the open mapping theorem using the continuity of linear differential operators w.r.t. the topology of $C^\infty(M, \mathbb{K})$ or of $\mathcal{D}(M, \mathbb{K})$ (beware that the latter is not Fréchet). This completes the sketch of proof of Theorem 3. □

Corollary 5. *Let P be a generalized d'Alembert operator on a globally hyperbolic spacetime M and $S \subset M$ be a spacelike Cauchy hypersurface in M with (timelike) unit normal v. Then for all $(f, u_0, u_1) \in C^\infty(M, \mathbb{C}) \oplus C^\infty(S, \mathbb{C}) \oplus C^\infty(S, \mathbb{C})$, there exists a unique $u \in C^\infty(M, \mathbb{C})$ solving (3.9). Moreover $\mathrm{supp}(u) \subset J_+^M(K) \cup J_-^M(K)$ with $K := \mathrm{supp}(u_0) \cup \mathrm{supp}(u_1) \cup \mathrm{supp}(f)$.*

Proof. Uniqueness already follows from Theorem 3. Let $(K_n)_n$ be a sequence of compact subsets of S with $K_n \subset \mathring{K}_{n+1}$ and $\cup_n K_n = S$. Identify M with $\mathbb{R} \times S$ and set $\widetilde{K}_n := D(\mathring{K}_n) \cap (]{-}n, n[\times S)$, where $D(\mathring{K}_n)$ is the so-called Cauchy development of $\mathring{K}_n$ in M, see Definition 11 on page 56. Then $(\widetilde{K}_n)_n$ is an increasing sequence of relatively compact and globally hyperbolic open subsets of M with $\cup_n \widetilde{K}_n = M$. Furthermore $\mathring{K}_n \subset S \simeq \{0\} \times S$ is a Cauchy hypersurface of $\widetilde{K}_n$ for every n. Let χ_n be a smooth function with compact support on M such that $\chi_{n|\widetilde{K}_n} = 1$. From Theorem 3 there exists a unique solution $v_n \in C^\infty(M, \mathbb{C})$ to the Cauchy problem

$$\left|\begin{array}{l} Pv_n = \chi_n f \\ v_{n|S} = \chi_n u_0 \\ \partial_v v_n = \chi_n u_1. \end{array}\right.$$

If $m > n$ then $v := v_m - v_n$ solves $Pv = 0$ on the globally hyperbolic manifold $\widetilde{K}_n$ with $v = \partial_v v = 0$ on the Cauchy hypersurface $\mathring{K}_n$ of $\widetilde{K}_n$; therefore, Theorem 3 implies $v = 0$ on $\widetilde{K}_n$. Hence $u(x) := v_n(x)$ for $x \in \widetilde{K}_n$ defines a smooth function u on M solving (3.9). The statement on the support is also a straightforward consequence of Theorem 3. This shows Corollary 5. □

3.5.4 Global Existence of Fundamental Solutions on Globally Hyperbolic Spacetimes

We come back to the issue of finding global fundamental solutions on globally hyperbolic spacetimes. Although they seem to be far away, Theorem 3 makes fundamental solutions easily accessible.

Theorem 4. *Let P be a generalized d'Alembert operator on a globally hyperbolic spacetime M. Then there exists for each $x \in M$ a unique fundamental solution*

$F_+(x)$ with past-compact support for P at x and a unique one $F_-(x)$ with future-compact support.

They satisfy

- $\mathrm{supp}(F_\pm(x)) \subset J_\pm^M(x)$ *and*
- *for every* $\varphi \in \mathcal{D}(M, \mathbb{C})$ *the map* $M \to \mathbb{C}$, $x \mapsto F_\pm(x)[\varphi]$ *is a smooth function with*

$$P^*(x \mapsto F_\pm(x)[\varphi]) = \varphi.$$

Proof. Uniqueness has already been obtained in Corollary 4, so that we just have to prove existence. Identify M with $\mathbb{R} \times S$, where $\partial/\partial t$ is future directed and each $\{s\} \times S$ is a smooth spacelike Cauchy hypersurface (this is always possible on globally hyperbolic spacetimes, see Theorem 1). Fix a smooth unit vector field ν normal to all $\{s\} \times S$. Set, for $x \in M$ and $\varphi \in \mathcal{D}(M, \mathbb{C})$,

$$F_+(x)[\varphi] := (\chi_\varphi)(x),$$

where χ_φ is the solution of the Cauchy problem

$$\begin{vmatrix} P^*\chi_\varphi & = \varphi \\ \chi_\varphi|_{\{t\}\times S} & = 0 \\ \partial_\nu \chi_\varphi|_{\{t\}\times S} & = 0 \end{vmatrix} \qquad (3.10)$$

and t is chosen such that $\mathrm{supp}(\varphi) \subset I_-^M(\{t\} \times S)$ (such a t can be found because of the compactness of $\mathrm{supp}(\varphi)$). If t is fixed then the existence of a solution of (3.10) is guaranteed by Theorem 3. However one has to show that χ_φ is well defined, i.e. does not depend on t. Let $t' \in \mathbb{R}$ be such that $\mathrm{supp}(\varphi) \subset I_-^M(\{t'\} \times S)$ and with, say, $t < t'$. Let $(\chi_\varphi)'$ be the solution of (3.10) with t' instead of t. We show that $(\chi_\varphi)'$ vanishes as well as its normal derivative on $\{t\} \times S$.

Since $\mathrm{supp}(\varphi)$ is compact there exists a $t_- < t$ such that $\mathrm{supp}(\varphi) \subset I_-^M(\{t_-\} \times S)$. Consider

$$M_{t_-} := \cup_{\tau > t_-} \{\tau\} \times S,$$

which is a globally hyperbolic spacetime in its own right and in which $\{t'\} \times S$ sits again as a Cauchy hypersurface. By assumption $\mathrm{supp}(\varphi)$ is contained in the complement of M_{t_-} in M, so that the restriction of $(\chi_\varphi)'$ onto M_{t_-} solves the Cauchy problem $Pu = 0$, $u|_{\{t'\}\times S} = 0$ and $\partial_\nu u|_{\{t'\}\times S} = 0$. The uniqueness of solutions (Theorem 3) implies that $(\chi_\varphi)'|_{M_{t_-}} = 0$, in particular $(\chi_\varphi)'$ vanishes in a neighbourhood of $\{t\} \times S$.

Now the smooth function $\chi := \chi_\varphi - (\chi_\varphi)'$ satisfies $P^*\chi = 0$ on M and $\chi|_{\{t\}\times S} = \partial_\nu \chi|_{\{t\}\times S} = 0$, hence by Theorem 3 again one concludes that $\chi = 0$ on M. Therefore χ_φ (and thus $F_+(\cdot)[\varphi]$) is well defined.

We next show that, for a fixed $x \in M$, the map $\varphi \mapsto F_+(x)[\varphi]$ is an advanced fundamental solution for P at x on M. The linearity as well as the continuity of $F_+(x)$ both directly follow from Theorem 3. On the other hand, given $\varphi \in \mathcal{D}(M, \mathbb{C})$, the function φ itself provides an obvious solution to $P^*u = P^*\varphi$ with $u_{|\{t\}\times S} = \partial_\nu u_{|\{t\}\times S} = 0$; since $\mathrm{supp}(P^*\varphi) \subset \mathrm{supp}(\varphi)$ is compact, Theorem 3 may be applied and we deduce that $\chi_{P^*\varphi} = \varphi$, which in turn implies from the definition of $F_+(x)$ that

$$
\begin{aligned}
P F_+(x)[\varphi] &= F_+(x)[P^*\varphi] \\
&= (\chi_{P^*\varphi})(x) \\
&= \varphi(x).
\end{aligned}
$$

This holds for all $\varphi \in \mathcal{D}(M, \mathbb{C})$, that is, $P F_+(x) = \delta_x$.

The support condition is equivalent to $\mathrm{supp}(\chi_\varphi) \subset J_-^M(\mathrm{supp}(\varphi))$ for every $\varphi \in \mathcal{D}(M, \mathbb{C})$. But for any such φ the open subset $M' := M \setminus J_-^M(\mathrm{supp}(\varphi))$ of M is again a globally hyperbolic manifold containing $\{t\} \times S$ as Cauchy hypersurface, where t is chosen as above (this follows from, e.g. [1, Lemma A.5.8] and a short reflection). The function $u := \chi_{\varphi|M'}$ satisfies $P^*u = 0$ with $u_{|\{t\}\times S} = \partial_\nu u_{|\{t\}\times S} = 0$; hence Theorem 3 again implies that $\chi_{\varphi|M'} = 0$, which was to be proved. Thus $F_+(x)$ is an advanced fundamental solution for P at x on M. That $x \mapsto F_+(x)[\varphi]$ is smooth with $P^*(x \mapsto F_+(x)[\varphi]) = \varphi$ for any $\varphi \in \mathcal{D}(M, \mathbb{C})$ is trivially seen from the definition of $F_+(\cdot)$. The construction of F_- is completely analogous, replacing all "+" by "−" and vice versa. This achieves the proof of Theorem 4.   □

In particular the wave equation $Pu = f$ with $f \in \mathcal{D}(M, \mathbb{C})$ possesses a unique solution $u_\pm \in C^\infty(M, \mathbb{C})$ with $\mathrm{supp}(u_\pm) \subset J_\pm^M(\mathrm{supp}(f))$; or equivalently with $\mathrm{supp}(u_+)$ (resp. $\mathrm{supp}(u_-)$) being past (resp. future) compact on a globally hyperbolic spacetime M.

Remark. Because of the definition of Riesz distributions we have only proved the existence of solutions to wave equations as well as fundamental solutions for $\mathbb{K} = \mathbb{C}$. In fact, all existence and uniqueness results from Corollary 1 to Theorem 4 still hold replacing $\mathbb{C}$ by $\mathbb{K} = \mathbb{R}$ for real valued generalized d'Alembert operators P, where "real-valued" means that $Pu \in C^\infty(M, \mathbb{R})$ whenever $u \in C^\infty(M, \mathbb{R})$ (or, equivalently, that the coefficients a_j and b_1 in local coordinates are real-valued functions, see Sect. 3.2.1). Indeed, we have already noticed in Sect. 3.3 that Riesz distributions for real parameters are real valued; moreover, if P is real valued then all objects involved in the local construction are real valued (e.g. the Riesz distributions $R_\pm^\Omega(2k+2, x)$ or the Hadamard coefficients, see Proposition 1), hence provide real-valued fundamental or classical solutions. In case uniqueness is available (such as in Theorems 3 and 4) the existence of real-valued solutions for such a P and for real-valued data straightforwardly follows from the corresponding result in the complex case, since the complex conjugate $\bar{u}$ of the distribution u then solves the same equation as u, hence $\bar{u} = u$.

3.6 Green's Operators

We now briefly sketch how solutions of wave equations can be encoded into a pair of operators, which furthermore offer an entrance door to (local) quantum field theory for generalized d'Alembert operators.

Definition 11. *Let P be a generalized d'Alembert operator on a spacetime M. A linear map*

$$G_+ : \mathcal{D}(M, \mathbb{K}) \longrightarrow C^\infty(M, \mathbb{K}),$$

satisfying

(i) $P \circ G_+ = \mathrm{Id}_{\mathcal{D}(M,\mathbb{K})}$,
(ii) $G_+ \circ P|_{\mathcal{D}(M,\mathbb{K})} = \mathrm{Id}_{\mathcal{D}(M,\mathbb{K})}$,
(iii) $\mathrm{supp}(G_+\varphi) \subset J_+^M(\mathrm{supp}(\varphi))$ *for all* $\varphi \in \mathcal{D}(M, \mathbb{K})$

is called advanced Green's operator *for P on M.*

A retarded Green's operator G_- for P on M is a linear map $\mathcal{D}(M, \mathbb{K}) \longrightarrow C^\infty(M, \mathbb{K})$ satisfying (i), (ii) and $\mathrm{supp}(G_-\varphi) \subset J_-^M(\mathrm{supp}(\varphi))$ for all $\varphi \in \mathcal{D}(M, \mathbb{K})$.

For P a (retarded or advanced) Green's operator is almost an inverse: it is a right inverse to P; however, a left inverse to $P|_{\mathcal{D}(M,\mathbb{K})}$ and not to P itself. This consideration reminds us of Sect. 3.2.2, where we have seen how fundamental solutions generally provide solutions to the corresponding wave equation for "every" right member. One could therefore expect a direct relationship between fundamental solutions and Green's operators. In fact Green's operators and fundamental solutions are two different versions of mainly the same concept.

Proposition 7. *Let P be a generalized d'Alembert operator on a spacetime M. Then advanced (resp. retarded) Green's operators for P stand in one-to-one correspondence with retarded (resp. advanced) fundamental solutions for P^*. More precisely, if there exists for every $x \in M$ a fundamental solution $F_\pm(x)$ for P^* at x on M with*

(a) $\mathrm{supp}(F_\pm(x)) \subset J_\pm^M(x)$,
(b) $x \mapsto F_\pm(x)[\varphi]$ *is smooth and*
(c) $P(x \mapsto F_\pm(x)[\varphi]) = \varphi$

for each $\varphi \in \mathcal{D}(M, \mathbb{K})$, then the formula

$$(G_\mp\varphi)(x) = F_\pm(x)[\varphi] \qquad \forall x \in M, \ \forall \varphi \in \mathcal{D}(M, \mathbb{K}) \tag{3.11}$$

defines a linear map $G_\mp : \mathcal{D}(M, \mathbb{K}) \to C^\infty(M, \mathbb{K})$ satisfying (i), (ii) in Definition 11 as well as $\mathrm{supp}(G_\mp\varphi) \subset J_\mp^M(\mathrm{supp}(\varphi))$.

Conversely, *every linear map $G_\mp : \mathcal{D}(M, \mathbb{K}) \to C^\infty(M, \mathbb{K})$ having those properties defines at each point $x \in M$ through (3.11) a fundamental solution $F_\pm(x)$ for P^* satisfying (a), (b) and (c).*

The proof of Proposition 7 is an easy exercise left to the reader. Combining it with Theorem 4, we obtain the following.

Corollary 6. *Every generalized d'Alembert operator on a globally hyperbolic space-time admits a unique advanced and a unique retarded Green's operator.*

We next list the properties that are needed for quantum field theory. On a given spacetime M we introduce the space

$$C_{sc}^{\infty}(M, \mathbb{K}) := \{ u \in C^{\infty}(M, \mathbb{K}) \,|\, \exists\, K \subset M \text{ compact s.t.}$$
$$\text{supp}(u) \subset J_+^M(K) \cup J_-^M(K) \}.$$

The "sc" stands for "spacelike compact", since in case M is globally hyperbolic the intersection of $J_+^M(K) \cup J_-^M(K)$ with any Cauchy hypersurface is compact. It can be proved that $C_{sc}^{\infty}(M, \mathbb{K})$ is a Fréchet vector space w.r.t. the topology for which a sequence $(u_j)_j$ converges towards 0 if and only if there exists a compact $K \subset M$ with $\text{supp}(u_j) \subset J_+^M(K) \cup J_-^M(K)$ for all j and such that $(u_j)_j$ converges to 0 in every C^k-norm on any compact subset of M.

Proposition 8. *Let P be a generalized d'Alembert operator on a spacetime M. Let G_+, G_- be advanced and retarded Green's operators for P on M. Set $G := G_+ - G_-$. Then the following holds:*

(i) The sequence

$$0 \longrightarrow \mathcal{D}(M, \mathbb{K}) \overset{P}{\longrightarrow} \mathcal{D}(M, \mathbb{K}) \overset{G}{\longrightarrow} C_{sc}^{\infty}(M, \mathbb{K}) \overset{P}{\longrightarrow} C_{sc}^{\infty}(M, \mathbb{K}) \qquad (3.12)$$

is a complex (i.e. the composition of any two successive maps is zero) which is exact at the first $\mathcal{D}(M, \mathbb{K})$.

(ii) If M is globally hyperbolic, then the formal adjoint of $G_{\pm}$ coincides with $G_{\mp}^*$, where G_+^*, G_-^* are Green's operators for P^* on M.

(iii) If M is globally hyperbolic, then the complex (3.12) is exact everywhere and all maps are sequentially continuous.

Proof. By definition of Green's operators, (3.12) is obviously a complex. Furthermore, if $\varphi \in \mathcal{D}(M, \mathbb{K})$ solves $P\varphi = 0$, then applying, e.g. G_+ one has $G_+(P\varphi) = \varphi = 0$, which shows exactness at the first $\mathcal{D}(M, \mathbb{K})$ and (i).

Assume now that M is globally hyperbolic. Let $\varphi, \psi \in \mathcal{D}(M, \mathbb{K})$, then $\text{supp}(G_{\pm}\varphi) \cap \text{supp}(G_{\mp}^*\psi)$ is compact, so that the following computation is justified:

$$\int_M \langle G_{\pm}\varphi, \psi \rangle dx = \int_M \langle G_{\pm}\varphi, P^* G_{\mp}^*(\psi) \rangle dx$$

$$= \int_M \langle P G_{\pm}\varphi, G_{\mp}^*\psi \rangle dx$$

$$= \int_M \langle \varphi, G_{\mp}^*\psi \rangle dx,$$

where $\langle\cdot,\cdot\rangle$ denotes the natural Euclidean or Hermitian inner product on $\mathbb{K}$. This proves (ii).

Still assuming M to be globally hyperbolic, let $\varphi \in \mathcal{D}(M,\mathbb{K})$ be such that $G\varphi = 0$. Then the function $\psi := G_+\varphi = G_-\varphi$ is smooth with $\mathrm{supp}(\psi) \subset J_+^M(\mathrm{supp}(\varphi)) \cap J_-^M(\mathrm{supp}(\varphi))$, which is compact. Moreover, $P\psi = PG_+\varphi = \varphi$. This shows exactness at the second $\mathcal{D}(M,\mathbb{K})$. Let now $u \in C_{\mathrm{sc}}^\infty(M,\mathbb{K})$ solve $Pu = 0$. After possibly enlarging K we may assume that a compact subset K of M exists such that $\mathrm{supp}(u) \subset I_+^M(K) \cup I_-^M(K)$. Let $\{\chi_+, \chi_-\}$ be a partition of unity subordinated to the open covering $\{I_+^M(K), I_-^M(K)\}$ of $I_+^M(K) \cup I_-^M(K)$. Setting $u_\pm := \chi_\pm u$ we obtain $u = u_+ + u_-$, where $u_\pm$ is smooth with $\mathrm{supp}(u_\pm) \subset I_\pm^M(K)$. Set now $\varphi := Pu_+ = -Pu_-$. It is a smooth function with $\mathrm{supp}(\varphi) \subset J_+^M(K) \cap J_-^M(K)$, which is compact, hence $\varphi \in \mathcal{D}(M,\mathbb{K})$. We check that $G\varphi = u$. Although $u_\pm$ does not have compact support, we may integrate $G_\pm\varphi$ against any $\psi \in \mathcal{D}(M,\mathbb{K})$; using the compacity of $\mathrm{supp}(u_\pm) \cap J_\mp^M(\mathrm{supp}(\psi))$ and (ii) we obtain

$$\int_M \langle G_\pm\varphi, \psi\rangle dx = \int_M \langle \varphi, (G_\pm)^*\psi\rangle dx$$
$$= \int_M \langle \varphi, G_\mp^*\psi\rangle dx$$
$$= \pm\int_M \langle Pu_\pm, G_\mp^*\psi\rangle dx$$
$$= \pm\int_M \langle u_\pm, P^*G_\mp^*\psi\rangle dx$$
$$= \pm\int_M \langle u_\pm, \psi\rangle dx,$$

that is, $G_\pm\varphi = \pm u_\pm$, so that $G\varphi = u_+ + u_- = u$. This shows exactness at the first $C_{\mathrm{sc}}^\infty(M,\mathbb{K})$. The sequential continuity of all maps of (3.12) follows from P being a differential operator and from Theorem 3. This shows (iii) and completes the proof of Proposition 8. □

One of the reasons why Green's operators are so important for quantum field theory is the following: Given a formally self-adjoint generalized d'Alembert operator P on a globally hyperbolic spacetime M, one can form a symplectic vector space in a canonical way, namely set

$$V := \mathcal{D}(M,\mathbb{K})/\mathrm{Ker}(G),$$

where $G := G_+ - G_-$ as above and G_+, G_- are Green's operators for P. From Proposition 8.(ii) the map $(\varphi, \psi) \mapsto \int_M \langle G\varphi, \psi\rangle dx$ defines a skew-symmetric bilinear form on V, which is by definition non-degenerate and hence a symplectic form on V. Now independently of this there also exists a canonical way to produce a C^*-algebra out of a symplectic vector space, which consists in defining its so-called CCR representation, where CCR stands for "canonical commutation relations". Composing both one obtains a kind of map – actually a functor – associating with each pair (M, P) a C^*-algebra. Of course this construction is made so as to translate into algebraic properties the analytical ones of the operator and the geometric ones of the underlying manifold; for example, an inclusion of manifolds corresponds to an inclusion of algebras (this has to do with functoriality) and if two globally hyperbolic open subsets of M are causally independent (i.e. if there is no causal curve from the closure of one to the closure of the other one) then the corresponding algebras commute. For more on quantization see the last chapter and [1, Chap. 4].

Acknowledgement The author would like to thank Christian Bär for his comments and his thorough reading of this survey.

References

1. Bär, C., Ginoux, N., Pfäffle, F.: Wave equations on Lorentzian manifolds and quantization. EMS Publishing House, Zürich (2007)
2. Lawson, H.B., Michelsohn, M.-L.: Spin Geometry. Princeton University Press, Princeton (1989)
3. Hörmander, L.: The analysis of linear partial differential operators, I–III. Springer-Verlag, Berlin (2007)

Chapter 4
Microlocal Analysis

Alexander Strohmaier

Microlocal Analysis deals with the singular behavior of distributions in phase space.

4.1 Introduction

Distributions appear in physics in various forms: as mass distributions of point particles, as Green's functions, and as propagators in quantum field theory. The theory of distributions not only has applications such as these in physics but is also widely and intensively used in mathematics, in particular in the theory of partial differential equations. For example, fundamental solutions to partial differential equations are usually singular distributions and the behavior of the singularities of these distributions encodes the behavior of the solutions. Microlocal analysis deals with the detailed analysis of such distributions. It turns out that the singularities of distributions can be localized in phase space. This leads to the notion of wavefront sets, a refinement of the notion of singular support. Physicists may find this natural: the singularities of solutions to partial differential equations are described by geometrical optics, and geometrical optics is equivalent to a classical system on classical phase space. In fact, some ideas used in microlocal analysis were developed in less rigorous form by physicists. Asymptotic expansions, the WKB approximation, transport equations, all well familiar to physicists, appear also in microlocal analysis, sometimes a little bit disguised.

In these notes I will follow an approach which is to a large extent due to Hörmander [1–3] and which aims at a precise and rigorous treatment of the propagation of singularities of distributions. The notes are essentially self-contained but some knowledge of the basic notions of functional analysis and topology is assumed. The material is organized as follows. In Sect. 4.2 I recall the theory of distributions and of Fourier transforms of distributions on an elementary level. In Sect. 4.3 I focus on the concept of the wavefront set as a refinement of the notion of the singular support. Sections 4.3.1 and 4.3.2 follow closely the treatment of Hörmander's book

A. Strohmaier (✉)
Department of Mathematical Sciences, Loughborough University, Leicestershire LE11 3TU, UK
e-mail: A.Strohmaier@lboro.ac.uk

Strohmaier, A.: *Microlocal Analysis*. Lect. Notes Phys. **786**, 85–127 (2009)
DOI 10.1007/978-3-642-02780-2_4   © Springer-Verlag Berlin Heidelberg 2009

[1, vol. I] and I discuss when a distribution may be pulled back under a smooth map and when two distributions may safely be multiplied. In Sect. 4.3.3 the wavefront sets of the fundamental solutions to the wave equation in Minkowski spacetime are calculated. In Sect. 4.3.4 the wavefront sets of the most commonly used scalar propagators of QFT in Minkowski spacetime are determined. Section 4.4 contains two fundamental results in microlocal analysis: microlocal elliptic regularity says in which direction solutions to partial differential equations may be singular and the propagation of singularity theorem clarifies how singularities of solutions to partial differential equations propagate. In Sect. 4.5 we demonstrate how these two results can be used to determine the wavefront sets of Green's distributions of any generalized d'Alembert operator on a globally hyperbolic spacetime. As realized by Radzikowski in his PhD thesis [4] the so-called Hadamard states, that are believed to be the physical states in QFT in curved spacetimes, can be characterized by their wavefront sets. I will give a brief explanation of this characterization in the end of these notes.

4.2 Distributions

4.2.1 Basic Definitions and Properties of Distributions

The so-called Dirac δ-function that is widely used in physics describes, for example, the density of a point mass: it is a positive function, has support at a single point, and it integrates to 1. It is an easy exercise to prove that there are no functions that satisfy the above properties. However, one can still calculate with this function and get reasonable results; and one can do this on a sound mathematical basis. It is the theory of distributions that achieves this.

In the following let $\mathcal{U}$ be a non-empty open subset of $\mathbb{R}^n$. The set of smooth complex-valued functions on $\mathcal{U}$ will be denoted by $\mathcal{E}(\mathcal{U}) = C^\infty(\mathcal{U})$ and the subset of functions with support compact in $\mathcal{U}$ will be denoted by $\mathcal{D}(\mathcal{U}) = C_0^\infty(\mathcal{U})$. Both spaces come equipped with topologies, namely $\mathcal{E}(\mathcal{U})$ is a Fréchet space with the family of semi-norms

$$p_{\alpha,K}(f) = \sup_{x \in K} |\partial^\alpha f(x)|. \tag{4.1}$$

where K runs over all compact subsets $K \subset \mathcal{U}$

This means a sequence of functions f_n converges to f in $\mathcal{E}(\mathcal{U})$ if and only if all its derivatives converge uniformly on compact subsets. Since any open set admits a countable exhaustion by compact subsets one can always replace the above family by an equivalent countable family of semi-norms. Thus, $\mathcal{E}(\mathcal{U})$ is indeed a Fréchet space.

The topology on $\mathcal{D}(\mathcal{U})$ is slightly more complicated to define. One chooses a compact exhaustion $K_1 \subset K_2 \subset, \ldots, \bigcup_n K_n = \mathcal{U}$. Then, let $\mathcal{D}(K_i)$ be the set of functions in $\mathcal{D}(\mathcal{U})$ with support in K_i. For each function in $\mathcal{D}(\mathcal{U})$ there exists an index k such that for all $n \geq k$ the function has support in K_n. In slightly different words this is the observation that $\mathcal{D}(\mathcal{U})$ is a direct limit of the system

$$\mathcal{D}(K_1) \hookrightarrow \mathcal{D}(K_1) \hookrightarrow \ldots . \qquad (4.2)$$

We endow $\mathcal{D}(K_i)$ with the relative topology as a subset of $\mathcal{E}(\mathcal{U})$, and $\mathcal{D}(\mathcal{U})$ with the inductive limit topology.

What this actually means is that a sequence f_n converges in $\mathcal{D}(\mathcal{U})$ to f if there exists a compact set K such that $\mathrm{supp}\, f_n \subset K$, $\mathrm{supp}\, f \subset K$, and all derivatives of f_n converge uniformly in K.

Definition 1. *The space $\mathcal{E}'(\mathcal{U})$ of compactly supported distributions on $\mathcal{U}$ is defined to be the topological dual of $\mathcal{E}(\mathcal{U})$, i.e., the space of continuous linear functionals on $\mathcal{E}(\mathcal{U})$.*

Let us see what it means for a functional on $\mathcal{E}(\mathcal{U})$ to be continuous.

Theorem 1. *A linear functional $\phi : \mathcal{E}(\mathcal{U}) \to \mathbb{C}$ is a distribution in $\mathcal{E}'(\mathcal{U})$ if and only if there is a compact set $K \subset U$, a $k \in \mathbb{N}$, and a positive constant C such that*

$$|\phi(f)| \leq C \sum_{|\alpha| \leq k} \sup_{x \in K} |\partial^\alpha f|$$

for all $f \in \mathcal{E}(\mathcal{U})$.

Proof. Obviously, if ϕ satisfies the above estimate then ϕ is continuous. On the other hand, if the above condition is not satisfied this means that given any compact set K and any k and C there is a function $f_{K,k,C}$ such that the inequality is wrong. For an exhaustion K_i let $C = k = i$. Then, there is a sequence f_i of functions such that

$$|\phi(f_i)| > i \sum_{|\alpha| \leq k} \sup_{x \in K_i} |\partial^\alpha f_i|.$$

The inequality in particular implies that $\phi(f_i) \neq 0$ and it does not change if f_i is multiplied by a constant. We can therefore re-scale the functions in such a way that $\phi(f_i) = 1$ and the above implies that $\sup_{K_i} |\partial^\alpha f_i| < 1/i$ for all $|\alpha| < i$. In particular, this means that f_i converges uniformly on all compact subsets to 0. Hence, ϕ is not continuous. $\qquad \square$

In the above the smallest integer k such that the inequality holds for some C and some compact K is called the order of the distribution.

Definition 2. *The space $\mathcal{D}'(\mathcal{U})$ of distributions on $\mathcal{U}$ is defined to be the topological dual of $\mathcal{D}(\mathcal{U})$, i.e., the space of continuous linear functionals on $\mathcal{D}(\mathcal{U})$.*

Similar to the above there is a criterion for a functional to be a distribution in $\mathcal{D}'$.

Theorem 2. *A linear functional* $\phi : \mathcal{D}(\mathcal{U}) \to \mathbb{C}$ *is a distribution in* $\mathcal{D}'(\mathcal{U})$ *if and only if for every compact set* $K \subset \mathcal{U}$ *there is a* $k \in \mathbb{N}$, *and a positive constant* C *such that*

$$|\phi(f)| \le C \sum_{|\alpha| \le k} \sup_K |\partial^\alpha f|$$

for all $f \in \mathcal{D}(K)$.

Proof. The estimate of course immediately implies continuity. Suppose conversely that the estimate does not hold. This means there exists a compact set $K \subset \mathcal{U}$ such that for every C, k there is a function $f_{C,k}$ such that the inequality does not hold. As above we can choose those functions in such a way that $\phi(f_{C,k}) = 1$. Again we can take $C = k = i$ and construct a sequence f_i of functions in $\mathcal{D}(K)$ such that

$$1 = |\phi(f_i)| > i \sum_{|\alpha| \le k} \sup_{x \in K} |\partial^\alpha f_i|,$$

$$\phi(f_i) = 1$$

and as above one gets $\sup_K |\partial^\alpha f_i| < 1/i$ for $|\alpha| < i$. In particular, this implies that f_i converges to 0 in $\mathcal{D}(\mathcal{U})$ and thus ϕ is not continuous.   □

The above two theorems may be (and are often) used to define distributions. And one may safely do this, since in the end it is those two inequalities that one checks if one wants to find out if a functional is a distribution. I use here the approach via locally convex topological vector spaces because it provides us with a language to express things in short terms.

If $\mathcal{U}' \subset \mathcal{U}$ is an open subset then $\mathcal{D}(\mathcal{U}')$ is a closed subspace of $\mathcal{D}(\mathcal{U})$ and there is a natural restriction map $\mathcal{D}'(\mathcal{U}) \to \mathcal{D}'(\mathcal{U}')$. We denote the restriction of a distribution ϕ to an open subset $\mathcal{U}'$ by $\phi|_{\mathcal{U}'}$.

Definition 3. *The support* supp ϕ *of a distribution* $\phi \in \mathcal{D}'(U)$ *is the smallest closed set* $\mathcal{O}$ *such that* $\phi|_{\mathcal{U} \setminus \mathcal{O}} = 0$.

Of course, every element in $\mathcal{E}'(\mathcal{U})$ is naturally an element in $\mathcal{D}'(\mathcal{U})$ simply because $\mathcal{D}(\mathcal{U})$ is continuously embedded into $\mathcal{E}(\mathcal{U})$. Such an element has compact support, since there exists a compact set K and $k \in \mathbb{N}$ and $C > 0$ such that

$$|\phi(f)| \le C \sum_{\alpha \le k} \sup |\partial^\alpha f| \qquad (4.3)$$

and therefore $\phi|_{\mathcal{U} \setminus K} = 0$. On the other hand, every element ϕ in $\mathcal{D}'(\mathcal{U})$ with compact support can be extended in a unique way to a functional in $\mathcal{E}(\mathcal{U})$ by choosing a smooth compactly supported function χ equal to 1 in a neighborhood of the support and defining the extension $\tilde{\phi} \in \mathcal{E}'(\mathcal{U})$ by

$$\tilde{\phi}(f) := \phi(\chi \cdot f). \qquad (4.4)$$

It is an easy exercise that multiplication by χ defines a continuous map from $\mathcal{E}(\mathcal{U})$ to $\mathcal{D}(\mathcal{U})$. Moreover, $\tilde{\phi}$ is independent of the choice of χ which is a simple consequence of the linearity of the functional.

In the following we will identify $\mathcal{E}'(\mathcal{U})$ with the space of compactly supported distributions in $\mathcal{D}'(\mathcal{U})$.

Now it is important to note that any locally integrable function $\phi \in L^1_{loc}(\mathcal{U})$ defines a distribution in $\mathcal{D}'(U)$ by

$$f \mapsto \int_{\mathcal{U}} f(x)\phi(x)dx.$$

The map $L^1_{loc}(\mathcal{U}) \to \mathcal{D}'(U)$ is injective, which means that the function is (up to a set of measure zero) determined by the distribution. In particular, every smooth function defines in this way a distribution and the support of a function as a distribution coincides with its support as a function. I would like to state the above-mentioned inclusions separately as

$$\mathcal{D}(\mathcal{U}) \subset \mathcal{E}'(\mathcal{U}) \subset \mathcal{D}'(\mathcal{U}),$$
$$\mathcal{E}(\mathcal{U}) \subset \mathcal{D}'(\mathcal{U}).$$

The most prominent example of a distribution that is not a function is the Dirac δ-distribution δ_{x_0} at a point $x_0 \in \mathcal{U}$. It is defined by

$$\delta_{x_0} : \mathcal{D}(\mathcal{U}) \to \mathbb{C}, \ f \mapsto f(x_0).$$

It is a zero-order distribution with support and singular support equal to $\{x_0\}$.

One now defines various operations with distributions in such a way that they coincide with the well-known operations on functions after restricting them.

Definition 4. *The product of a distribution ϕ in $\mathcal{D}'(\mathcal{U})$ by a function $g \in \mathcal{E}(U)$ is the distribution in $\mathcal{D}'(\mathcal{U})$ defined by*

$$(g \cdot \phi)(f) := \phi(g \cdot f). \tag{4.5}$$

This is certainly just the ordinary point-wise product in case ϕ is a function:

$$\int (g(x)\phi(x)) \, f(x)dx = \int \phi(x) \, (g(x)f(x)) \, dx.$$

For example the product of δ_{x_0} by a function $g \in \mathcal{E}(\mathcal{U})$ is by definition given by

$$(g \cdot \delta_{x_0})(f) = \delta_{x_0}(g \cdot f) = g(x_0)f(x_0) = g(x_0)\delta_{x_0}(f)$$

and therefore $g \, \delta_{x_0} = g(x_0)\delta_{x_0}$.

Now the derivative of a distribution can also be defined imitating the case when ϕ is a function namely if $\phi \in \mathcal{E}(\mathcal{U})$ and $f \in \mathcal{D}(\mathcal{U})$ one can partially integrate and obtain

$$(\partial^\alpha \phi)(f) = \int \partial^\alpha \phi(x) f(x) dx$$

$$= (-1)^{|\alpha|} \int \phi(x) \partial^\alpha f(x) dx = (-1)^{|\alpha|} \phi(\partial^\alpha f).$$

This justifies the following definition.

Definition 5. *The partial derivatives $\partial^\alpha \phi$ of $\phi \in \mathcal{D}'(\mathcal{U})$ are defined by*

$$(\partial^\alpha \phi)(f) = (-1)^{|\alpha|} \phi(\partial^\alpha f).$$

More generally for a differential operator P define

$$(P\phi)(f) = \phi(P^t f),$$

where P^t is the formal transpose operator.

And we can see an interesting phenomenon here. Any distribution can be arbitrarily often differentiated and the result will again be a distribution. For example any function in $L^1_{loc}(\mathcal{U})$ has distributional derivatives of any order. As an example one may consider the distributional derivative of the Heaviside step function $H(x)$ on $\mathbb{R}$ which is 1 for $x \geq 0$ and 0 for $x < 0$. Its distributional derivative is the δ-function at 0 because

$$H'(f) = \int H(x) \left(-f'(x)\right) dx = -\int_0^\infty f'(x) dx = f(0) = \delta_0(f)$$

and this justifies $H' = \delta_0$.

Another prominent example is the equation $\Delta(1/r) = 4\pi \delta_0$ in $\mathbb{R}^3$ which is justified by

$$\left(\Delta \frac{1}{r}\right)(f) = \int \frac{1}{r} \Delta f(x) dx = \lim_{\epsilon \to 0} \int_{\mathbb{R}^3 \backslash B_\epsilon} \frac{1}{r} \Delta f(x) dx$$

$$= \lim_{\epsilon \to 0} \left(\int_{\mathbb{R}^3 \backslash B_\epsilon} (\Delta \frac{1}{r}) f(x) dx + \int_{\partial B_\epsilon} \frac{1}{r} \frac{\partial f}{\partial r} dx + \int_{\partial B_\epsilon} \frac{1}{r^2} f(x) dx \right)$$

$$= \lim_{\epsilon \to 0} \left(4\pi \epsilon^2 \frac{1}{\epsilon} \frac{\partial f}{\partial r} |_{r=0} + 4\pi \epsilon^2 \frac{1}{\epsilon^2} f(0) \right) = 4\pi f(0) = 4\pi \delta_0(f),$$

where we have used the second Green's identity and the fact that $1/r$ is harmonic away from 0. This formula makes a lot of sense because indeed the potential $1/r$ is that of a point mass at 0 and it shows how powerful the concept of distributions already in such a simple situation is.

The following definition provides us with a language to specify where a distribution is smooth and where it is singular.

Definition 6. *The singular support* singsupp ϕ *of $\phi \in \mathcal{D}'(U)$ is the smallest closed subset $\mathcal{O}$ such that $\phi|_{\mathcal{U} \backslash \mathcal{O}} \in \mathcal{E}(\mathcal{U} \backslash \mathcal{O})$.*

4.2.2 Convolution and Approximation by Smooth Functions

The Dirac δ-distribution can be approximated weakly by smooth compactly supported functions. For example if we have an arbitrary function $\phi \in \mathcal{D}(\mathbb{R}^n)$ with

$$\phi(x) > 0,$$
$$\phi(0) \neq 0,$$
$$\int \phi(x)dx = 1,$$

then the rescaled function $\phi_\lambda(x) = \lambda^n \phi(\lambda x)$ with $\lambda \in \mathbb{R}_+$ is also compactly supported and smooth and satisfies the above properties. As $\lambda \to \infty$ we get supp $\phi_\lambda \to \{0\}$. This is already enough to conclude that

$$\lim_{n \to \infty} \phi_n(f) = \lim_{n \to \infty} \int \phi_n(x) f(x)dx = f(0) = \delta_0(f),$$

which is easily shown by splitting the integral into supp ϕ_n and $\mathbb{R}^n \setminus$ supp ϕ_n and using the continuity of f. A family of functions ϕ_λ satisfying the above conditions and supp $\phi_\lambda \to \{0\}$ is called a delta family .

The weak-* topologies in $\mathcal{D}'$ and $\mathcal{E}'$, respectively, are the topologies of point-wise convergence: a net (or a sequence) ψ_α of distributions converges to ϕ if and only if $\phi_\alpha(f)$ converges to $\phi(f)$ for every test function f in $\mathcal{D}$ or $\mathcal{E}$, respectively. The above shows that there is a sequence of compactly supported smooth functions that converges in the weak-* topology to δ_0. As it turns out any distribution can be approximated by test functions. In order to prove this we need the concept of convolution of a distribution with a function. Remember that the convolution of two functions f and g on $\mathbb{R}^n$ is defined as the function

$$(f * g)(x) := \int f(y)g(x - y)dy,$$

whenever this expression is defined. This is certainly the case if one function is integrable while the other is bounded. Note that here we used the group structure of $\mathbb{R}^n$ and the convolution depends on this structure. One can make sense of this for distributions as well in the usual manner by just imitating.

Definition 7. *Let $\phi \in \mathcal{D}'(\mathbb{R}^n)$ and $f \in \mathcal{D}(\mathbb{R}^n)$. Then, the convolution*

$$\phi * f \in C^\infty(\mathbb{R}^n)$$

is defined by $(\phi * f)(x) = \phi(\tilde{f}_x)$, where $\tilde{f}_x(y) = f(x - y)$.

It is immediate from the Taylor expansion of f that indeed $\phi * f$ is a function in $C^\infty(\mathbb{R}^n)$ and its derivative is given by

$$\partial^\alpha(\phi * f)(x) = \phi * (\partial^\alpha f). \tag{4.6}$$

A consequence one gets that for a delta family ϕ_λ as above the sequence $f * \phi_n = \phi_n * f$ converges to f in $\mathcal{D}(\mathbb{R}^n)$ for any $f \in \mathcal{D}(\mathbb{R}^n)$ since we can interchange convolution and differentiation.

Theorem 3. *Let $\phi \in \mathcal{D}'(\mathbb{R}^n)$ and $f, g \in \mathcal{D}(\mathbb{R}^n)$. Then,*

- $(\phi * f) * g = \phi * (f * g)$,
- $\operatorname{supp} \phi * f \subset \operatorname{supp} \phi + \operatorname{supp} f$,
- $\partial^\alpha(\phi * f) = (\partial^\alpha\phi) * f = \phi * (\partial^\alpha f)$.

Proof. By definition we have

$$\phi * (f * g)(x) = \phi\left(\int f(x - t - \cdot)g(t)dt\right) = \phi\left(\int f(t - \cdot)g(x - t)dt\right).$$

We need to argue why

$$\phi\left(\int f(t - \cdot)g(x - t)dt\right) = \int \phi(f(t - \cdot))\, g(x - t)dt.$$

This seems obvious since the integral is linear. It needs to be checked, however, that the error as one approximates the integral by a Riemann sum tends to zero in the topology of $\mathcal{D}$ for fixed x as the mesh goes to zero. It is not difficult to see that this is indeed the case. It follows from the fact that f and g have both compact support and all integrals converge absolutely. This implies that all derivatives are again Riemann sums which converge uniformly. For a function g denoted by $\tilde{g}$ the reflected function $\tilde{g}(x) = g(-x)$. Then, clearly,

$$\phi(g) = (\phi * \tilde{g})(0).$$

Now the statement relating the supports is obviously true if ϕ is an integrable function because the integrand in the convolution integral vanishes for x which is not in $\operatorname{supp} \phi + \operatorname{supp} f$. The more general statement for distributions follows immediately from this and the equation

$$(\phi * f)(g) = ((\phi * f) * \tilde{g})(0) = (\phi * (f * \tilde{g}))(0) = \phi(\tilde{f} * g),$$

using the definition of the support of a distribution. The third formula follows directly from Eqn. (4.6). □

And the main statement in this section is the promised density result.

Theorem 4. *$\mathcal{D}(\mathcal{U})$ is sequentially dense in $\mathcal{E}'(\mathcal{U})$ and also in $\mathcal{D}'(\mathcal{U})$. The above definitions of multiplication by a function, partial differentials of distributions, and convolutions are the unique continuous extensions from $\mathcal{D}(\mathcal{U})$.*

Proof. Let us first show the statement for $\mathcal{E}'(\mathcal{U})$. Choose a δ-family ϕ_λ as above. Next choose a sequence of functions χ_n in $\mathcal{D}(\mathcal{U})$ such that for every compact subset $K \subset \mathcal{U}$ there is an N with $\chi_n = 1$ on K for all $n > N$. Then, $\chi_n \psi$ is a sequence of distributions in $\mathcal{E}'(\mathbb{R}^n)$ with support in K. Now one can match those sequences in such a way that the convolution $(\chi_n \psi) * \phi_n$ has support in $\mathcal{U}$. This is achieved if the radius of the ball which contains the support of ϕ_n is smaller than the distance of K from $\mathbb{R}^n \backslash \mathcal{U}$. I claim that $(\chi_n \psi) * \phi_n$ is a sequence of compactly supported smooth functions that converges to ψ in the weak-* topology. The functions are smooth and compactly supported by construction. We only need to check the convergence:

$$\lim_{n \to \infty} ((\chi_n \psi) * \phi_n)(f) = \lim_{n \to \infty} (\chi_n \psi) * (\phi_n * \tilde{f})(0)$$

$$= \lim_{n \to \infty} (\chi_n \psi)(\tilde{\phi}_n * f) = \lim_{n \to \infty} \psi(\tilde{\phi}_n * f),$$

where we have used in the last step that χ_n is one in a neighborhood of the support of $\tilde{\phi}_n * f$ for large enough n. Clearly, $\tilde{\phi}_n$ is also a δ-family and therefore $\tilde{\phi}_n * f$ converges in $\mathcal{D}$ to f. Thus, we conclude

$$\lim_{n \to \infty} ((\chi_n \psi) * \phi_n)(f) = \psi(f).$$

The statement for $\mathcal{E}'$ follows from that for $\mathcal{D}'$. $\qquad\square$

4.2.3 Schwartz Distributions

Although the emphasis here is clearly on distributions in open subsets of $\mathbb{R}^n$ and later on manifolds it is very convenient to introduce a third space of distributions which is very special for $\mathbb{R}^n$ and which is related to the Fourier transform. We say a smooth function f is in $\mathcal{S}(\mathbb{R}^n)$ if

$$\sup_x |x^\alpha \partial^\beta f(x)| < \infty$$

for all multi-indices α and β. This space is then obviously a Fréchet space with semi-norms given by

$$p_{\alpha,\beta}(f) = \sup_x |x^\alpha \partial^\beta f(x)|.$$

The space of Schwartz distributions $\mathcal{S}'(\mathbb{R}^n)$ (or tempered distributions) is then defined as the topological dual of this space. In the same way as before one shows that a linear functional ϕ on $\mathcal{S}(\mathbb{R}^n)$ is in $\mathcal{S}'(\mathbb{R}^n)$ if and only if there is an $N > 0$ and a constant $C > 0$ such that

$$|\phi(f)| \leq C \sum_{|\alpha|+|\beta| \leq N} \sup_x |x^\alpha \partial^\beta f(x)|.$$

Exercise 1. Give a proof of this.

To get some idea of Schwartz functions let us prove some statements about them.

Lemma 1. *For any polynomial P in n variables the maps*

$$P(\partial) : S(\mathbb{R}^n) \to S(\mathbb{R}^n), f \mapsto P(\partial_x)f,$$
$$P(x) : S(\mathbb{R}^n) \to S(\mathbb{R}^n), f \mapsto P(x) \cdot f$$

are continuous. Moreover, the map

$$f \mapsto \int f(x)dx$$

is a continuous linear functional on $S(\mathbb{R}^n)$.

Proof. We only have to show that the maps are bounded. For the first statement this follows from the obvious inequality

$$p_{\alpha,\beta}(P(\partial_x)f) = \sup_x |x^\alpha \partial_x^\beta P(\partial_x)f| \le C_{\alpha,\beta} \sum_{|\alpha'|+|\beta'|+\deg(P)\le|\alpha|+|\beta|} \sup_x |x^{\alpha'} \partial_x^{\beta'} f|.$$

The second statement follows from the same formula

$$p_{\alpha,\beta}(P(x)f) = \sup_x |x^\alpha \partial_x^\beta P(x)f| \le C_{\alpha,\beta} \sum_{|\alpha'|+|\beta'|+\deg(P)\le|\alpha|+|\beta|} \sup_x |x^{\alpha'} \partial_x^{\beta'} f|,$$

which is obtained after using the Leibniz rule to interchange the order in which $P(x)$ and the derivatives are applied. The fact that the integral is continuous follows from

$$\left| \int f(x)dx \right| = \left| \int (1 + |x|)^{-n-1}(1 + |x|)^{n+1} f(x)dx \right|$$

$$\le \int (1 + |x|)^{-n-1}|(1 + |x|)^{n+1} f(x)|dx$$

$$\le \left(\int (1 + |x|)^{-n-1}dx \right) |\sup_x |(1 + |x|)^{n+1} f(x)|$$

$$\le C \sup_x |(1 + |x|)^{n+1} f(x)|,$$

and the right-hand side is bounded by $\sum_{|\alpha|+|\beta|\le n+1} p_{\alpha,\beta}(f)$. $\square$

Of course, any distribution in $\mathcal{E}'(\mathcal{U})$ is automatically a Schwartz distribution. Functions that are bounded by a polynomial are also examples of Schwartz distributions that are not necessarily in $\mathcal{E}'(\mathbb{R}^n)$. For any open subset of $\mathbb{R}^n$ we have the following inclusions:

$$\mathcal{E}'(\mathcal{U}) \subset \mathcal{S}'(\mathbb{R}^n) \subset \mathcal{D}'(\mathbb{R}^n).$$

Here is a small remark for physicists. The Schwartz spaces are important since they treat momentum and position equally. They arise naturally in the theory of the harmonic oscillator as domains of smoothness of the operator $\Delta + |x|^2$.

4.2.4 The Fourier Transform

The Fourier transform $\hat{f}$ of a function in $f \in L^1(\mathbb{R}^n)$ is defined as

$$\hat{f}(\xi) = (2\pi)^{-\frac{n}{2}} \int_{\mathbb{R}^n} f(x) e^{-i\langle \xi, x \rangle} dx.$$

This function is bounded by the L^1-norm of f since

$$|\hat{f}(\xi)| = (2\pi)^{-\frac{n}{2}} |\int_{\mathbb{R}^n} f(x) e^{-i\langle \xi, x \rangle} dx| \le (2\pi)^{-\frac{n}{2}} \int_{\mathbb{R}^n} |f(x)| dx = (2\pi)^{-\frac{n}{2}} \|f\|_1.$$

I would like to state here the most important properties of the Fourier transform which are well covered in standard textbooks on analysis.

Proposition 1. *The following statements hold:*

(i) *if $f, g \in L^1(\mathbb{R}^n)$, then $\widehat{f * g}(\xi) = (2\pi)^{\frac{n}{2}} \hat{f}(\xi)\hat{g}(\xi)$;*
(ii) *if $f \in L^1(\mathbb{R}^n) \cap L^2(\mathbb{R}^n)$ then $\hat{f} \in L^2(\mathbb{R}^n)$ and the Plancherel formula $\|\hat{f}\|_2 = \|f\|_2$ holds;*
(iii) *the Fourier transform extends to a unitary map $L^2(\mathbb{R}^n) \to L^2(\mathbb{R}^n)$;*
(iv) *for $f \in L^2(\mathbb{R}^n)$ the inverse of the Fourier transform $f^\vee$ is given by $f^\vee(x) = \hat{f}(-x)$.*

Of course, every function in $\mathcal{S}(\mathbb{R}^n)$ is in $L^1(\mathbb{R}^n)$ and therefore, the Fourier function of a Schwartz function is well defined. The following is a standard result, I would like to prove it here, since the method of proof will be applied later to more complicated situations and it is important to understand it.

Proposition 2. *The Fourier transform of a function in $\mathcal{S}(\mathbb{R}^n)$ is again in $\mathcal{S}(\mathbb{R}^n)$. The map $\mathcal{F} : \mathcal{S}(\mathbb{R}^n) \to \mathcal{S}(\mathbb{R}^n)$, $f \mapsto \hat{f}$ is a linear continuous bijection.*

Proof. Since each function in $f \in \mathcal{S}(\mathbb{R}^n)$ decays faster than any negative power this implies that for any multi-index α the function $x^\alpha f(x)$ is absolutely integrable. This proves that $\hat{f}$ is smooth and its derivatives are given by

$$\partial_\xi^\alpha \hat{f}(\xi) = (2\pi)^{-\frac{n}{2}} \int_{\mathbb{R}^n} f(x) \partial_\xi^\alpha e^{-i\langle \xi, x \rangle} dx$$

$$= (2\pi)^{-\frac{n}{2}} \int_{\mathbb{R}^n} f(x)(-ix)^\alpha e^{-i\langle \xi, x \rangle} dx = \widehat{(-ix)^\alpha f}.$$

In the same way one gets

$$\xi^\alpha \partial_\xi^\beta \hat{f}(\xi) = (2\pi)^{-\frac{n}{2}} \int_{\mathbb{R}^n} f(x)(-ix)^\beta \xi^\alpha e^{-i\langle \xi, x \rangle} dx$$

$$= (2\pi)^{-\frac{n}{2}} \int_{\mathbb{R}^n} f(x)(-ix)^\beta (i\partial_x)^\alpha e^{-i\langle \xi, x \rangle} dx$$

$$= (2\pi)^{-\frac{n}{2}} \int_{\mathbb{R}^n} f(x)(-ix)^\beta (i\partial_x)^\alpha e^{-i\langle \xi, x \rangle} dx$$

and after integration by parts

$$\xi^\alpha \partial_\xi^\beta \hat{f}(\xi) = (2\pi)^{-\frac{n}{2}} \int_{\mathbb{R}^n} (i\partial_x)^\alpha \left((-ix)^\beta f(x) \right) e^{-i\langle \xi, x \rangle} dx.$$

If $f \in \mathcal{S}(\mathbb{R}^n)$ then the right-hand side is finite for all α and β. Moreover,

$$|\xi^\alpha \partial_\xi^\beta \hat{f}(\xi)| \le (2\pi)^{-\frac{n}{2}} |\int_{\mathbb{R}^n} (i\partial_x)^\alpha \left((-ix)^\beta f(x) \right) dx|.$$

The right-hand side does not depend to ξ anymore and by Lemma 1 the map

$$f \mapsto \int_{\mathbb{R}^n} (i\partial_x)^\alpha \left((-ix)^\beta f(x) \right) dx$$

is continuous. Therefore, also $\mathcal{F}$ is continuous. The same is true for the inverse transform and we conclude that the Fourier transform is a continuous bijection.   □

Here again the remark for physicists. As I said earlier $\mathcal{S}(\mathbb{R}^n)$ can be equivalently defined as the domain of smoothness of the operator $\Delta + |x|^2$, which describes the harmonic oscillator. Since the Fourier transform interchanges the operators Δ and $|x|^2$ it leaves the operator $\Delta + |x|^2$ invariant. From this it is obvious that the domain of smoothness is left invariant as well. Moreover, the equivalent family of semi-norms $p_k(f) := \|(\Delta + |x|^2)^k f\|_{L^2}$ is invariant under the Fourier transform. Thus, the map is continuous.

By duality (using the Plancherel formula) the Fourier transform extends to a weak-* continuous linear map

$$\mathcal{F} : \mathcal{S}'(\mathbb{R}^n) \to \mathcal{S}'(\mathbb{R}^n)$$

simply by defining $(\mathcal{F}\phi)(f) := \phi(\hat{f})$.

By the proof of Proposition 2 and duality the following is immediate.

Proposition 3. *For $\phi \in S'(\mathbb{R}^n)$ the following formulae hold:*

$$\widehat{\partial_x^\alpha \phi}(\xi) = (i\xi)^\alpha \hat{\phi}(\xi),$$
$$\widehat{x^\alpha \phi}(\xi) = (i\partial_\xi)^\alpha \hat{\phi}(\xi).$$

Since every distribution with compact support $\phi \in \mathcal{E}'(\mathcal{U})$ is also in $S'(\mathbb{R}^n)$ the Fourier transform makes sense for such a distribution. Of course, for distributions the situation is much simpler from the beginning. We could have defined the Fourier transform of such a distribution directly because such a distribution may be paired with any smooth function.

Theorem 5. *For $\phi \in \mathcal{E}'(\mathcal{U})$ the Fourier transform $\hat{\phi}$ is the smooth function given by*

$$\hat{\phi}(\xi) = \phi(e_\xi),$$

where e_ξ is the smooth function $e_\xi(x) = (2\pi)^{-n/2} e^{-i(\xi,x)}$. The function $\hat{\phi}$ extends to an entire function $\hat{\phi}(z)$ such that $\hat{\phi}$ is of uniform exponential type, i.e., there is a $k \in \mathbb{N}$, a $C_1 > 0$, and a $C_2 > 0$ such that

$$|\hat{\phi}(z)| \le C_1 (1 + |z|)^k e^{C_2 |\Im z|}.$$

In particular, $\hat{\phi}$ is polynomially bounded on $\mathbb{R}$.

Proof. Suppose that $g \in S(\mathbb{R}^n)$. By the same argument as in the proof of Theorem 3 the integral

$$\int g(\xi) e_\xi \, d\xi = \hat{g}$$

can be approximated by a Riemann sum and the error converges to zero in $S(\mathbb{R}^n)$ as the mesh goes to 0. Therefore,

$$\int \phi(e_\xi) g(\xi) d\xi = \phi\left(\int e_\xi g(\xi) d\xi\right) = \phi(\hat{g}),$$

which shows that indeed the Fourier transform defined by duality is the smooth function given by $\phi(e_\xi)$. Since ϕ may be paired with any smooth function this expression makes sense for any $\xi \in \mathbb{C}$. Moreover, e_z depends holomorphically on z and Taylors theorem implies immediately that $\phi(e_z)$ is complex differentiable with complex differential equal to $\phi(\frac{d}{dz} e_z)$. The estimate follows from

$$|\phi(e_z)| \le C \sum_{|\alpha| \le k} \sup_{x \in K} |\partial^\alpha e_z(x)|$$

(see Theorem 1) since

$$|\partial^\alpha e_z(x)| = |z^\alpha e_z(x)| \le (1 + |z|)^{|\alpha|} e^{|x||\Im(z)|}. \qquad \square$$

4.3 Singularities of Distributions and the Wavefront Set

It is an old wisdom in physics that small scales in space correspond to large scales in momentum space. Already the Heisenberg uncertainty relation states that the localization of a particle requires a high momentum. In CERN resolving detailed structures of particles requires extremely high energies. Singularities of distributions are local phenomena. What happens to singular distributions if we Fourier transform them? Can we see their singular behavior in momentum space? Let us start with the following simple observation.

Theorem 6. $\phi \in \mathcal{E}'(\mathcal{U})$ *is smooth if and only if for every N there is a constant C_N such that*

$$|\hat{\phi}(\xi)| \le C_N (1 + |\xi|)^{-N}.$$

Proof. If such a distribution is smooth it means that $\phi \in \mathcal{D}(\mathcal{U})$ and in particular that $\phi \in \mathcal{S}(\mathbb{R}^n)$. Therefore, $\hat{\phi} \in \mathcal{S}(\mathbb{R}^n)$ which implies the estimate. Conversely, suppose the estimate holds. Since $\hat{\phi}(\xi)$ decays faster than any polynomial

$$\frac{1}{(2\pi)^{n/2}} \int \hat{\phi}(\xi) P(\partial_x) e^{i\langle \xi, x \rangle} d\xi \qquad (4.7)$$

is well defined for all polynomials P. Therefore, the above defines a smooth function in L^2. By the inversion theorem this function coincides with ϕ as a distribution. $\square$

Therefore, what we have actually shown is that the singular support of a distribution $\phi \in \mathcal{D}'(\mathcal{U})$ is the complement of the set of points $x \in \mathcal{U}$ such that there is a function $f \in \mathcal{D}(\mathcal{U})$ with $f(x) = 1$ such that $\widehat{f \cdot \phi}$ is rapidly decaying. Let us microlocalize this. There might be directions where $\widehat{f \cdot \phi}$ is decaying and others where it is not. This motivates the following definition of the wavefront set which is a kind of microlocal singular support. I will give the definition first and will give a more detailed explanation later.

Definition 8. *For a distribution $\phi \in \mathcal{D}'(\mathcal{U})$ the wavefront set $\mathrm{WF}(\phi)$ is the complement in $\mathcal{U} \times \mathbb{R}^n \backslash \{0\}$ of the set of points $(x, \xi) \in \mathcal{U} \times \mathbb{R}^n \backslash \{0\}$ such that there exist*

- *a function $f \in \mathcal{D}(\mathcal{U})$ with $f(x) = 1$,*
- *an open conic neighborhood Γ of ξ, with*

$$\sup_{\xi \in \Gamma} (1 + |\xi|)^N |\widehat{f \cdot \phi}(\xi)| < \infty$$

for all $N \in \mathbb{N}_0$.

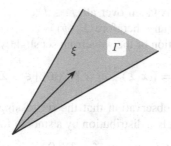

Fig. 4.1 A conic neighborhood of ξ

An open conic neighborhood is by definition an open neighborhood which is invariant under the action of $\mathbb{R}_+$ by multiplication. This means that it is of the form

$$\{\lambda x \mid x \in S, \lambda \in \mathbb{R}_+\},$$

where S is an open subset of the sphere S^{n-1} (see Figure 4.1). Similarly, a closed conic neighborhood is one of the form $\mathbb{R}_+ S$, where S is a closed subset of S^{n-1}.

The first remark I would like to make is that from the definition it is clear that WF(ϕ) is a closed subset of $\mathcal{U} \times \mathbb{R}^n \setminus \{0\}$, because the condition for a point to be in its complement is open. To more deeply understand this definition let us look at the set of singular directions of a distribution $\phi \in \mathcal{E}'(\mathcal{U})$.

Definition 9. *For a compactly supported distribution $\phi \in \mathcal{E}'(\mathcal{U})$ we say a point $\xi \in \mathbb{R}^n \setminus \{0\}$ is regular directed if there is an open conic neighborhood Γ of ξ, with*

$$\sup_{\xi' \in \Gamma} (1 + |\xi'|)^N |\widehat{\phi}(\xi')| < \infty$$

for all $N \in \mathbb{N}_0$.

This means roughly that the set of regular directions is the set of directions in which the Fourier transform has a rapid decay. Note, however, that the set of regular directed points is defined in such a way that it is an open set, since we require decay in a whole conic neighborhood. Now the set $\Sigma(\phi)$, the set of singular directions, is by definition the complement of the set of regular directed points:

$$\Sigma(\phi) = \{\xi \in \mathbb{R}^n \setminus \{0\} \mid \xi \text{ is not regular directed}\}.$$

Again this is roughly the set of directions in which the Fourier transform does not decay rapidly, thus it is those directions that give rise to singularities. Now for any distribution $\phi \in \mathcal{D}'(\mathcal{U})$ and a point $x \in \mathcal{U}$ one may define

$$\Sigma_x(\phi) = \bigcap_f \Sigma(f\phi),$$

where the intersection is taken over all $f \in \mathcal{D}(\mathcal{U})$ with $f(x) \neq 0$. This is the intersection of closed sets and therefore $\Sigma_x(\phi)$ is a closed subset of $\mathbb{R}^n \backslash \{0\}$. Now we can rephrase the definition of the wavefront set slightly, because

$$\mathrm{WF}(\phi) = \{(x, \xi) \in \mathcal{U} \times \mathbb{R}^n \backslash \{0\} \mid \xi \in \Sigma_x(\phi)\}. \tag{4.8}$$

It is a very important observation that the set of singular directions does not become larger if we multiply a distribution by a smooth function.

Lemma 2. *For any $\phi \in \mathcal{E}'(\mathcal{U})$ and any $f \in \mathcal{D}(\mathcal{U})$ we have*

$$\Sigma(f\phi) \subset \Sigma(\phi).$$

Proof. Since ϕ has compact support $\hat{\phi}$ is a polynomially bounded function, i.e., there exists $N > 0$ and $C > 0$ such that

$$|\hat{\phi}(\xi)| \leq C(1 + |\xi|)^N.$$

Since f is smooth and compactly supported it is in $\mathcal{S}(\mathbb{R}^n)$ and therefore also $\hat{f} \in \mathcal{S}(\mathbb{R}^n)$. This means in particular that $\hat{f}$ is rapidly decaying in all directions. The Fourier transform of the product $f\phi$ is given by a convolution

$$\widehat{f\phi}(\xi) = (2\pi)^{-\frac{n}{2}} \int_{\mathbb{R}^n} \hat{\phi}(\xi - \eta)\hat{f}(\eta)d\xi.$$

Now suppose that ξ_0 is a regular directed point for $\hat{\phi}$. This means that there is a conic neighborhood Γ of ξ_0 where the $\hat{\phi}$ has rapid decay. We can of course choose a smaller open cone Γ' which is a cone over a small open ball around ξ_0. Now for every $\xi \in \Gamma'$ we can look at the open ball $B(\xi)$ of maximal radius $R(\xi)$ that is contained in Γ and of course as $|\xi|$ goes to infinity the radius $R(\xi)$ is bounded from below by $c|\xi|$ for some $c > 0$ as a consequence of the intercept theorem (see Fig. 4.2).

This means if the convolution integral is split into two parts

$$\widehat{f\phi}(\xi) = (2\pi)^{-\frac{n}{2}} \int_{\mathbb{R}^n} \hat{\phi}(\xi - \eta)\hat{f}(\eta)d\xi$$

$$= (2\pi)^{-\frac{n}{2}} \int_{|\eta| \leq R(\xi)} \hat{\phi}(\xi - \eta)\hat{f}(\eta)d\xi + (2\pi)^{-\frac{n}{2}}$$

$$\times \int_{|\eta| \geq R(\xi)} \hat{\phi}(\xi - \eta)\hat{f}(\eta)d\xi,$$

then in the first integral we have $|\xi - \eta| \geq c'|\xi|$ for some $c' > 0$ and moreover $\xi - \eta$ is contained in Γ. But $\hat{f}$ is a bounded function and in Γ by assumption $\hat{\phi}$ is rapidly decaying. Therefore, the first integral is rapidly decaying as $|\xi| \to \infty$ in Γ'. In the second integral we have $|\eta| \geq c|\xi|$. Since $\hat{\phi}$ is bounded by a polynomial and $\hat{f}$ is

Fig. 4.2 The cones Γ and Γ'

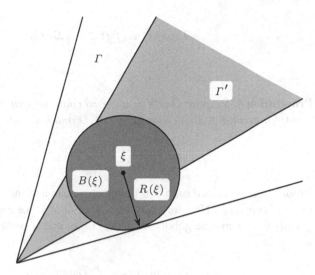

rapidly decaying we conclude that also the second integral is rapidly decaying as $|\xi| \to \infty$ in Γ'.   □

Note that this implies that the definition of WF is a local one because $\Sigma_x(f\phi) = \Sigma_x(\phi)$ if $f(x) \neq 0$. This can be seen immediately from

$$\Sigma_x(f\phi) = \bigcap_{g, g(x) \neq 0} \Sigma(fg\phi) \subset \bigcap_{g, g(x) \neq 0} \Sigma(g\phi) = \Sigma_x(\phi),$$

$$\Sigma_x(\phi) = \bigcap_{g, g(x) \neq 0} \Sigma(g\phi) \subset \bigcap_{g, g(x) \neq 0, g = f\tilde{g}} \Sigma(g\phi) = \bigcap_{g, g(x) \neq 0} \Sigma(gf\phi) = \Sigma_x(f\phi).$$

Moreover, if $f(x) \neq 0$ and supp $f \to \{x\}$ we get $\Sigma(f\phi) \to \Sigma_x(\phi)$. To be more precise:

Lemma 3. *For every open neighborhood V of $\Sigma_x(\phi)$ there is a neighborhood $\mathcal{U}_x$ of x such that*

$$\Sigma(f\phi) \subset V$$

for all f with supp $(f) \subset \mathcal{U}_x$.

Proof. The intersection of the unit sphere with the complement of V is compact. Moreover, the complements of $\Sigma(f\phi)$ form an open cover of this set. Therefore, there is a finite sub-cover. This means that there are finitely many $g_1, \ldots, g_k$ with $g_i(x) \neq 0$ and $\bigcap_{i=1}^{k} \Sigma(g_i\phi) \subset V$. Now choose $\mathcal{U}_x$ in such a way that all g_i are nonzero there. Then, for every $f \in \mathcal{D}(\mathcal{U}_x)$ also $\tilde{f} = f(g_1 g_2 \cdots g_k)^{-1} \in \mathcal{D}(\mathcal{U}_x)$. Thus, every g_i is a factor of f and consequently

$$\Sigma_x(\phi) \subset \Sigma(fu) \subset \bigcap_{i=1}^{k} \Sigma(g_i\phi) \subset V.$$

$\square$

Proposition 4. *Suppose* $O \times V$ *is a closed conic set that does not intersect* $\mathrm{WF}(\phi)$. *Then, for every* $f \in \mathcal{D}(\mathcal{U})$ *with* supp $f \subset O$ *and* $k \in \mathbb{N}_0$:

$$\sup_{\xi \in V}(1 + |\xi|)^k \widehat{f\phi}(\xi) < \infty.$$

Proof. Note that since supp f is compact we can assume without loss of generality that O is compact. Now for every (x_0, ξ_0) in $O \times V$ we can find a positive function g with $g = 1$ in a neighborhood $\mathcal{U}(x_0)$ of x_0 and a conic neighborhood Γ_{ξ_0} of ξ_0 such that

$$\sup_{\xi \in \Gamma}(1 + |\xi|)^k |\widehat{f\phi}(\xi)| < \infty.$$

Of course, $\mathcal{U}(x_0) \times \Gamma_{\xi_0}$ covers $O \times V$. Since $O \times (V \cap S^{n-1})$ is compact there is a finite sub-cover. This implies that

$$\sup_{\xi \in V}(1 + |\xi|)^k |\widehat{h\phi}(\xi)| < \infty,$$

where h is the sum of the functions g corresponding to the finite sub-cover. Since h is greater than or equal to 1 on the support of f we can multiply by fh^{-1} and obtain $\Sigma(f\phi) \cap V = \emptyset$. By the same argument as above by compactness of $V \cap S^{n-1}$ there are uniform estimates and we finally obtain

$$\sup_{\xi \in V}(1 + |\xi|)^k \widehat{f\phi}(\xi) < \infty.$$

$\square$

This means that the set $\Sigma_x(\phi)$ can be computed by calculating the Fourier transform of $f\phi$ and localize f more and more at x. Of course then also the wavefront set can be computed in this way. The intuition is that the wavefront set does not only say where a distribution is singular but also tells us in which direction it is singular.

Theorem 7. *The projection of* $\mathrm{WF}(\phi)$ *onto the first factor equals* singsupp ϕ. *Moreover*, $\mathrm{WF}(\phi) \subset \mathrm{WF}(f\phi)$.

Proof. Let us first show that $\mathrm{pr}_1(\mathrm{WF}(\phi))$ equals singsupp ϕ. If x is a point where ϕ is smooth then obviously we can find f with $f(x) \neq 0$ such that $f\phi$ is smooth and then $\Sigma_x(\phi) = \emptyset$. Conversely, suppose that $\Sigma_x(\phi) = \emptyset$. Then, by the above we can find a neighborhood $\mathcal{U}_x$ such that $\Sigma(f\phi) = \emptyset$ for all $f \in \mathcal{D}(\mathcal{U}_x)$. In particular, we may choose f to be 1 in a smaller neighborhood $\tilde{U} \subset \mathcal{U}_x$ of x. This means that

$f\phi$ is smooth and agrees with ϕ on $\tilde{U}$. Therefore, ϕ is regular at x. The statement $\mathrm{WF}(\phi) \subset \mathrm{WF}(f\phi)$ is obvious from the definition since

$$\Sigma_x(\phi) \subset \Sigma_x(f\phi).$$

□

Theorem 8. *For any $\phi \in \mathcal{D}'(\mathcal{U})$ we have $\mathrm{WF}(\partial^\alpha \phi) \subset \mathrm{WF}(\phi)$.*

Proof. The proof of this involves a trick which is very often used to localize statements in microlocal analysis. Suppose that (x_0, ξ_0) were not in the wavefront set of ϕ. In other words, there is a function $f \in \mathcal{D}(\mathcal{U})$ with $f(x) \neq 0$ with $\xi_0 \notin \Sigma(f\phi)$. We already saw that f can be chosen to be one in a neighborhood of 1. Now choose $g \in \mathcal{D}(\mathcal{U})$ such that $f = 1$ on the support of g and such that $g(x_0) \neq 0$. Since ∂^α is local

$$g\partial^\alpha f\phi = g\partial^\alpha \phi.$$

Then, $\xi_0 \notin \Sigma(f\phi)$ implies $\xi_0 \notin \Sigma(\partial^\alpha f\phi)$ since the partial differentiation becomes multiplication by $(i\xi)^\alpha$ after Fourier transformation. This again implies $\xi_0 \notin \Sigma(g\partial^\alpha f\phi) = \Sigma(g\partial^\alpha \phi)$. Therefore, (x_0, ξ_0) is a regular directed point as well for $\partial^\alpha \phi$.   □

Note that there is a more compact way of giving the same argument

$$\Sigma_x(\partial^\alpha \phi) \subset \Sigma(g\partial^\alpha u) = \Sigma(g\partial^\alpha f u) \subset \Sigma(\partial^\alpha f u) \subset \Sigma(f\phi)$$

and now just let supp f go to x.

Before we investigate closer the properties of wavefront sets I would like to give examples. Let us calculate the wavefront set of the Dirac δ-distribution at 0 in $\mathbb{R}^n$. First, the support of the distribution is $\{0\}$ and therefore we need to concentrate only on $\Sigma_0(\delta_0)$. Now for any f we get

$$\widehat{f\delta_0} = \frac{1}{\sqrt{2\pi}} f(0),$$

which is not decaying in any direction. Therefore,

$$\mathrm{WF}(\delta_0) = \{0\} \times (\mathbb{R}^n \backslash \{0\}).$$

Another example is the distribution $F = \lim_{\epsilon \to +0} 1/(x + i\epsilon)$ defined by

$$F(f) := \lim_{\epsilon \to +0} \int_{\mathbb{R}} \frac{1}{x + i\epsilon} f(x)dx.$$

The Fourier transform of $1/(x + i\epsilon)$ for $\epsilon > 0$ can easily be calculated using the residue theorem

$$\frac{1}{\sqrt{2\pi}} \int_{\mathbb{R}} \frac{1}{x+i\epsilon} e^{-i\xi x} dx = -\sqrt{2\pi} i H(\xi) e^{-\xi\epsilon}$$

and therefore $\hat{F} = -\sqrt{2\pi} i H(\xi)$ which at the same time shows that F is indeed a Schwartz distribution. Moreover, the Fourier transform of fF can be directly evaluated

$$\widehat{fF}(\xi) = \frac{1}{\sqrt{2\pi}}(\hat{f} * \hat{F})(\xi) = -i \int_{-\infty}^{\xi} \hat{f}(\eta)d\eta,$$

which decays rapidly as $\xi \to -\infty$ and tends to $f(0)$ as $\xi \to \infty$. Therefore,

$$\mathrm{WF}(F) = \{0\} \times \mathbb{R}^+.$$

Of course, we have the formula

$$\lim_{\epsilon\to+0} \left(\frac{1}{x+i\epsilon} - \frac{1}{x-i\epsilon} \right) = -2\pi i \delta_0$$

and we realize that we can split the δ-distribution as a sum of two distributions, one with wavefront equal to $\{0\} \times \mathbb{R}_+$, the other one with wavefront equal to $\{0\} \times \mathbb{R}_-$. This splitting is achieved here by splitting the Fourier transform into two functions with support $\mathbb{R}_+$ and $\mathbb{R}_-$, respectively.

Here is another example that shows a connection with energy in quantum theories. Suppose that $\mathcal{H}$ is a Hilbert space and H a self-adjoint operator. Denote by $U(t) = e^{iHt}$ the strongly continuous one-parameter group guaranteed to exist by Stones theorem. Then, for any two vectors ϕ and ψ the function

$$F_t = \langle \phi, U_t \psi \rangle$$

is continuous and therefore defines a distribution in $\mathcal{D}'(\mathbb{R})$. If $H > 0$ then the wavefront set of this distribution is contained in $\mathbb{R} \times \mathbb{R}_+$ because for $g \in \mathcal{D}(\mathbb{R})$ we have by spectral calculus

$$\widehat{gF} = g * \langle \phi, dE\psi \rangle,$$

where dE is the spectral measure interpreted as an operator-valued distribution. If H is positive the spectral measure has support in $\mathbb{R}_+$ and therefore $\widehat{gF}(t)$ decays rapidly as $t \longrightarrow -\infty$.

Exercise 2. Compute the wavefront set of a smooth measure on a linear subspace $S \subset \mathbb{R}^n$. More precisely, let $g \in C^\infty(S)$ and let ϕ be the distribution defined by

$$\phi(f) = \int_S f(x)g(x)dx,$$

where integration is over S.

For the observation that $WF(\phi + \psi) \subset WF(\phi) \cup WF(\psi)$ one obtains that differential operators do not increase the wavefront set.

Corollary 1. *For any partial differential operator P we have*

$$\mathrm{WF}(P\phi) \subset \mathrm{WF}(\phi).$$

4.3.1 Hörmander's Topology

For later purposes we will consider the space of distributions $\mathcal{D}'_\Gamma(\mathcal{U})$ for some closed cone $\Gamma \subset \mathcal{U} \times (\mathbb{R}^n \backslash \{0\})$. This is the set of distributions $\phi \in \mathcal{D}'(\mathcal{U})$ with $\mathrm{WF}(\phi) \subset \Gamma$.

Suppose that $\phi \in \mathcal{D}'_\Gamma(\mathcal{U})$ and f is a smooth function with compact support and V a closed cone in $\mathbb{R}^n$ such that

$$(\mathrm{supp}\, f \times V) \cap \Gamma = \emptyset.$$

Then, by Proposition 4

$$\sup_{\xi \in V}(1 + |\xi|)^k |\widehat{f\phi}| < \infty.$$

Conversely, if the above is true for all f and V with $(\mathrm{supp}\, f \times V) \cap \Gamma = \emptyset$, it follows from the definition of the wavefront set that Γ does not intersect $\mathrm{WF}(\phi)$. It is therefore natural to use the above constants as additional semi-norms for a topology on $\mathcal{D}'_\Gamma(\mathcal{U})$. Thus, the topology on $\mathcal{D}'_\Gamma(\mathcal{U})$ is defined as the locally convex topology defined by the following set of semi-norms:

$$p_f(\phi) = |\phi(f)|, \quad f \in \mathcal{D}(\mathcal{U}),$$
$$p_{f,V,k}(\phi) = \sup_{\xi \in V}(1 + |\xi|)^k |\widehat{f\phi}(\xi)|, \quad k \in \mathbb{N}_0, f \in \mathcal{D}(\mathcal{U}) : (\mathrm{supp}\,(f) \times V) \cap \Gamma = \emptyset,$$

where in the second line the index set is the set of all (f, V, k) with $k \in N$, $V \in \mathbb{R}^n \backslash \{0\}$ a closed cone, and $f \in \mathcal{D}(\mathcal{U})$ such that

$$(\mathrm{supp}\,(f) \times V) \cap \Gamma = \emptyset.$$

In the first line the index set is the set of all $f \in \mathcal{D}(\mathcal{U})$, thus the topology defined in this way is stronger than the weak-* topology.

Now the same way as in the proof of Theorem 4, namely by convolution with a δ-family one can show that $\mathcal{D}(\mathcal{U})$ is sequentially dense in $\mathcal{D}'_\Gamma(\mathcal{U})$ for every Γ.

4.3.2 Covariance

The aim of this section is to show that the wavefront set of a distribution transforms nicely under a change of coordinates. This makes wavefront sets important, since it will allow us to define them on manifolds. In order to see how the wavefront set transforms we need to study how the Fourier transform behaves under a change of coordinates. If F is a smooth map from $\mathcal{U}_1 \subset \mathbb{R}^n$ to $\mathcal{U}_2 \subset \mathbb{R}^m$ and ϕ is a function in $\mathcal{D}(\mathcal{U}_2)$ then the pull back $F^*\phi \in \mathcal{E}(\mathcal{U}_1)$ is defined by $F^*\phi(x) = \phi(F(x))$. In case the map is not properly supported the pullback $F^*\phi$ is in general not compactly supported. We can, however, localize it with some test function $\chi \in \mathcal{D}(\mathcal{U}_1)$ and look at the maps

$$F_\chi^* : \phi \mapsto \chi \cdot (F^*\phi).$$

The F_χ^* encode all the properties of the pullback map and of course, choosing a partition of unity, the pullback can be reconstructed from them. Since we want to analyze when we can pull back singular functions it is natural to ask how this map looks like in Fourier space. Using the inversion formula for the Fourier transform we have

$$F_\chi^*(\phi)(x) = \chi(x)\phi(F(x)) = \frac{1}{(2\pi)^{\frac{n}{2}}} \int \chi(x)\hat{\phi}(\xi)e^{iF(x)\xi} d\xi$$

and therefore

$$\widehat{F_\chi^*(\phi)}(\xi) = \frac{1}{(2\pi)^n} \int \hat{\phi}(\eta)T_\chi(\eta, \xi)d\eta,$$

$$T_\chi(\eta, \xi) = \int \chi(x)e^{i(\langle F(x),\eta\rangle - \langle x,\xi\rangle)}dx.$$

This means that the pullback F_χ^* is in Fourier space implemented by the integral operator T_χ. If we want to see for which distributions a pullback can be defined and how the singularities transform we need to see how this integral kernel behaves at infinity. Such oscillatory integrals like the one that defines this integral kernel appear very often in microlocal analysis. Their decay properties are encoded in the behavior of the phase functions as the following lemma shows.

Lemma 4. *Suppose* $u \in \mathcal{D}(\mathcal{U})$ *and* $f \in \mathcal{E}(\mathcal{U})$ *real valued such that* $f'(x) \neq 0$ *on* supp u. *Then,*

$$\omega^k \left| \int u(x)e^{i\omega f(x)}dx \right| \leq C_{k,f} \|u\|_{C^k} \sup_x |f'(x)|^{-2k}$$

and $C_{k,f}$ *can be chosen bounded for all* f *in a bounded subset of* C^k.

Proof. The idea is the same as in the proof of Proposition 2 for the Fourier transform: integration by parts. Define a differential operator

$$L = \omega^{-1} \left(\sum_{i=1}^{n} |\frac{\partial f}{\partial x_i}|^2 \right)^{-1} \sum_{k} \frac{\partial f}{\partial x_k} \frac{\partial}{\partial x_k}.$$

Then, it follows by direct calculation that

$$L e^{i\omega f(x)} = e^{i\omega f(x)}.$$

Therefore,

$$\int u(x) e^{i\omega f(x)} dx = \int u(x) L^k e^{i\omega f(x)} dx = \int \left((L^*)^k u(x) \right) e^{i\omega f(x)} dx,$$

where L^* is the formal adjoint operator

$$L^* u = -\sum_{k} \frac{\partial}{\partial x_k} \left(\omega^{-1} \left(\sum_{i=1}^{n} |\frac{\partial f}{\partial x_i}|^2 \right)^{-1} \frac{\partial f}{\partial x_k} u \right).$$

The iterate $(L^*)^k u$ can be evaluated using the Leibniz rule and the highest powers of $|f'(x)|^{-1}$ that appear after differentiation are $|f'(x)|^{-2k}$. Moreover, the derivatives of u enter in a linear manner and the highest derivatives appearing are those of order k. Additional derivatives of f enter in a polynomial manner and the highest derivative appearing is of order k. Thus, for f in a bounded subset of C^k the constants $C_{k,f}$ can be chosen to be bounded for fixed k. □

Note that this is not the strongest statement one can get. One can use the Cauchy–Schwarz inequality and the Leibniz rule to get a sharper estimate (see Theorem 7.7.1 in [1, Vol. I]). For our purposes the above is sufficient and I shall therefore not try to optimize constants. Since our integral operator $T_\chi(\eta, \xi)$ has the above form we can apply this result to get an estimate on its decay properties. If we apply Theorem 4 to our integral kernel we get the following corollary.

Corollary 2. *The following estimate holds for* $T_\chi(\eta, \xi)$:

$$|T_\chi(\eta, \xi)| \leq C_{\chi,k}(1 + |\eta|)^k (1 + |\xi|)^{-k}.$$

Moreover, if V and W are two closed conic subsets of $\mathbb{R}^n \setminus \{0\}$ such that $d F_x^*(\eta) \neq \xi$ for all $(\eta, \xi) \in V \times W$ and $x \in \text{supp}(\chi)$, then

$$|T_\chi(\eta, \xi)| \leq C_{\chi,V,W,k}(1 + |\eta| + |\xi|)^{-k}, \quad (\eta, \xi) \in V \times W.$$

Proof. The first inequality follows from Theorem 4 by setting $\omega = |\xi|$, $u(x) = \chi(x) e^{i\langle F(x), \xi \rangle}$ and $f(x) = |\xi|^{-1}\langle \xi, x \rangle$. Since $\|df\| = 1$ and f is bounded in C^k as ξ varies one gets

$$|T_\chi(\eta, \xi)| \leq C' \|\chi(x) e^{i\langle F(x), \eta \rangle}\|_{C^k} (1 + |\xi|)^{-k}.$$

Using the Leibniz rule one gets

$$\|\chi(x)e^{i\langle F(x),\eta\rangle}\|_{C^k} \leq C''(1+|\eta|)^k,$$

which proves the first inequality. The second inequality follows from Theorem 4 by setting $u(x) = \chi(x)$ and $f(x) = (|\eta|+|\xi|)^{-1}(\langle \eta, F(x)\rangle - \langle \xi, x\rangle)$ and $\omega = |\eta|+|\xi|$. Since

$$df = (|\eta|+|\xi|)^{-1}(\langle \eta, dF_x\rangle - \langle \xi, dx\rangle) = (|\eta|+|\xi|)^{-1}(\langle dF_x^*(\eta) - \xi, dx\rangle)$$

the assumptions imply that $\|df\| \geq \epsilon > 0$ for $(\eta, \xi) \in V \times W$ and $x \in \text{supp}(\chi)$. Again, f remains bounded in C^k as η and ξ vary. Thus, one gets

$$|T_\chi(\eta,\xi)| \leq C'''(1+|\eta|+|\xi|)^{-k},$$

which proves the second inequality. $\quad\square$

First recall that the normal set N_F of the map F is defined by

$$N_F := \{(F(x), \xi) \in U_2 \times \mathbb{R}^n \mid (dF^*)(\xi) = 0\}.$$

Theorem 9. *Let Γ be a closed cone in $U_2 \times (\mathbb{R}^n\setminus\{0\})$ and let $F : U_1 \to U_2$ be a smooth map such that $N_F \cap \Gamma = \emptyset$. Then, the pullback of functions $F^* : \mathcal{E}(U_2) \to \mathcal{E}(U_1)$ has a unique sequentially continuous extension to a sequentially continuous map $\mathcal{D}'_\Gamma(U_2) \to \mathcal{D}'(U_1)$ (which we for simplicity also denote by F^*).*

Proof. The proof is not difficult but quite involved notationally. That is why I would like to split it into three steps.
Step 1: The first step is to show that the problem is actually local. Let g_α be some partition of unity on U_1 and assume that f_α is a family of compactly supported functions on U_2 which are equal to 1 on $F(\text{supp } g_\alpha)$. Then,

$$F^*(\phi) = \sum_\alpha g_\alpha F^*(f_\alpha\phi).$$

It is therefore enough to find sequentially continuous extensions of the maps $\phi \mapsto g_\alpha F^*(f_\alpha\phi)$ for some partition of unity. Suppose that x_0 is a point in U_1. Since we assumed that $N_F \cap \Gamma = \emptyset$ we can choose a compact neighborhood K of $F(x_0)$ and an open neighborhood $\mathcal{O}$ of x such that $\overline{F(\mathcal{O})} \subset \text{int}(K)$ such that the following condition holds: There is an $\epsilon > 0$ such that the closure V of the set $\cup_{x\in\mathcal{O}}\{\xi | dF_x^*(\xi) \leq \epsilon\}$ satisfies $(K \times V) \cap \Gamma = \emptyset$. The such chosen maps define a cover of U_1 and we can choose a locally finite refinement which allows for a partition of unity as above. This shows that it is enough to find a sequentially continuous extensions of

$$(F|_{\mathcal{O}_\alpha})^* : C_0^\infty(K_\alpha) \to C^\infty(\mathcal{O}_\alpha)$$

to $\mathcal{D}'_\Gamma(K_\alpha)$ where we can assume that there is a closed cone $V_\alpha \subset \mathbb{R}^n$ with $(K \times V_\alpha) \cap \Gamma = \emptyset$ and $dF_x^*(\xi) > \epsilon > 0$ for all $x \in \mathcal{O}_\alpha$ and $\xi \notin V_\alpha$.

Step 2: We know from step 1 that it is enough to look at the restrictions of the pullback map to the described open sets. We want to show that the formula

$$\langle F^*(\phi), \chi \rangle = \int F_\chi^*(\phi)(x)dx = (2\pi)^{n/2}\widehat{F_\chi^*(\phi)}(0)$$

$$= \frac{1}{(2\pi)^{n/2}} \int \hat{\phi}(\eta)T_\chi(\eta, 0)d\eta$$

for supp $\chi \in \mathcal{O}_\alpha$ can be used to define an extension of the pullback to the set $\mathcal{D}'_\Gamma(K_\alpha)$. Suppose that supp $(\chi) \subset \mathcal{O}_\alpha$. We can split the integral over an integral over V and its complement

$$\langle F^*(\phi), \chi \rangle = \frac{1}{(2\pi)^{n/2}} \int \hat{\phi}(\eta)T_\chi(\eta, 0)d\eta$$

$$= \frac{1}{(2\pi)^{n/2}} \int_V \hat{\phi}(\eta)T_\chi(\eta, 0)d\eta$$

$$+ \frac{1}{(2\pi)^{n/2}} \int_{\mathbb{R}^n \setminus V} \hat{\phi}(\eta)T_\chi(\eta, 0)d\eta.$$

Both integrals make sense for distributions $\phi \in \mathcal{D}'_\Gamma(K_\alpha)$: the first integral converges because on V the function $\hat{\phi}(\eta)$ decays rapidly, whereas $|T_\chi(\eta, 0)| \leq \int |\chi(x)|dx$. This also implies that it depends continuously on χ. The second integral converges because for $\xi \notin V$ we have $dF_x^*(\xi) > \epsilon$ and therefore Corollary 2 implies that

$$|T_\chi(\eta, 0)| \leq C_k \|\chi\|_{C^k}(1 + |\eta|)^{-k},$$

whereas $\hat{\phi}(\eta)$ is polynomially bounded, i.e.,

$$|\hat{\phi}(\eta)| \leq C(1 + |\eta|)^M$$

for some M. Hence,

$$|\int_{\mathbb{R}^n \setminus V} \hat{\phi}(\eta)T_\chi(\eta, 0)d\eta| \leq C\|\chi\|_{C^{M+n+1}} \int (1 + |\eta|)^{-n-1}d\eta \leq C'\|\chi\|_{C^{M+n+1}}.$$

Therefore, also the second integral depends continuously on χ and defines a distribution.

Step 3: In a last step we need to show that the above defines indeed a sequentially continuous extension of the usual pullback. For this it is enough to show that the above two integrals depend sequentially continuously on $\phi \in \mathcal{D}'_\Gamma(K)$. Suppose that there is a sequence ϕ_k of distributions in $\mathcal{D}'_\Gamma(K)$ converging to 0. This means in particular that

$$\sup_{\eta \in V} |\hat{\phi}_k(\eta)|(1 + |\eta|)^N \to 0$$

for all $N > 0$. In particular, the first integral can be bounded by

$$\sup_{\eta \in V} |\hat{\phi}_k(\eta)|(1 + |\eta|)^{n+1}$$

and it thus converges to zero. Since $\mathcal{D}(K)$ is a Fréchet space we can apply the uniform boundedness principle. Since ϕ_k converges weakly the sequence $\phi_k(f)$ is bounded and by the uniform boundedness principle

$$|\hat{\phi}_k(\eta)| \leq C(1 + |\eta|)^M.$$

Thus, the integrand in the second integral is bounded by an L^1-function and by the dominated convergence theorem we can interchange limit and integration. Since ϕ_k converges weakly to zero $\hat{\phi}$ converges to zero point-wise. Therefore, also the second integral converges to zero.   □

In particular, the pullback is always defined for submersions. Let us denote by dF^* the pullback of co-vectors, so that for a subset of $\mathcal{O} \subset \mathcal{U}_2 \times \mathbb{R}^n$ we have by definition

$$dF^*(\mathcal{O}) = \{(x, dF_x^*(\xi)) \in \mathcal{U}_1 \times \mathbb{R}^n \mid (F(x), \xi) \in \mathcal{O}\}.$$

It is now natural to ask what the wavefront set of a pulled back distribution is.

Theorem 10. *Suppose that* $N_F \cap \Gamma = \emptyset$. *Then, the pullback* $F^*\phi$ *of a distribution in* $\mathcal{D}_\Gamma'(\mathcal{U}_2)$ *is in* $\mathcal{D}_{dF^*(\Gamma)}'(\mathcal{U}_1)$ *and the map*

$$F^* : \mathcal{D}_\Gamma'(\mathcal{U}_2) \to \mathcal{D}_{dF^*(\Gamma)}'(\mathcal{U}_1)$$

is sequentially continuous.

Proof. **Step 1:** For any point $\{x_0\} \in \mathcal{U}_1$ and closed cone $W \in \mathbb{R}^n$ such that $x_0 \times W \cap dF^*(\Gamma) = \emptyset$ there is a compact neighborhood K of $F(x_0)$ and an open neighborhood $\mathcal{O}$ of x_0 such that $\overline{F(\mathcal{O})} \subset \text{int}(K)$ such that the following condition holds: there is an $\epsilon > 0$ such that

$$\|dF_x^*(\eta) - \xi\| > \epsilon$$

for all $x \in \mathcal{O}$, $\eta \in W$, and $(x', \xi) \in \Gamma$ for some $x' \in K$. If we show that for $\phi \in \mathcal{D}_\Gamma'(K)$ the pullback $F^*(\phi)$ satisfies $\Sigma(F^*(\phi)) \cap W = \emptyset$, then we can use the same partition of unity argument as in the proof of the previous theorem to show that $\text{WF}(\phi) \subset dF^*(\Gamma)$ for any $\phi \in \mathcal{D}_\Gamma'(\mathcal{U}_2)$. Moreover, the continuity statement reduces to the statement that if the sequence ϕ_k converges to 0 in $\mathcal{D}_\Gamma'(K)$ then

$$\sup_{\xi \in W} |\widehat{F^*(\phi_k)}(\xi)|(1 + |\xi|)^N$$

converges to 0 for every N and $\chi \in \mathcal{D}(\mathcal{O})$ with $\xi(x_0) = 1$.

Step 2: Suppose $K, \mathcal{O}$, and W satisfy the conditions in step 1. Let $\chi \in \mathcal{D}(\mathcal{O})$ with $\chi(x_0) = 1$. For $\phi \in \mathcal{D}'_\Gamma(K)$ the Fourier transform of $\chi F^*(\phi)$ is given by

$$\widehat{F^*(\chi\phi)}(\xi) = \frac{1}{(2\pi)^n} \int \hat{\phi}(\eta) T_\chi(\eta, \xi) d\eta.$$

Let V be a small enough open neighborhood of

$$\{\eta \in \mathbb{R}^n \mid (x', \eta) \in \Gamma \text{ for some } x' \in K\}$$

such that $dF_x^*(\eta) \neq \xi$ for any $(\eta, \xi) \in V \times W$ and $x \in \text{supp}(\chi)$. A direct application of Corollary 2 to the split integral gives

$$\sup_{\xi \in W} |\widehat{F^*(\phi_k)}(\xi)|(1 + |\xi|)^N$$

$$\leq C_L \int_V \hat{\phi}(\eta)(1 + |\xi| + |\eta|)^{-L+N} d\eta$$

$$+ C' \int_{\mathbb{R}^n \setminus V} \hat{\phi}(\eta)(1 + |\eta|)^N d\eta.$$

Since $\hat{\phi}$ is polynomially bounded we can for any N choose L large enough so that the right-hand side is finite. Consequently $\{x_0\} \times W$ does not intersect $\text{WF}(F^*(\phi))$.

Step 3: It remains to show sequential continuity. This is now the same argument as in the proof of the previous theorem. If ϕ_k converges to 0 in $\mathcal{D}'_\Gamma(K)$ then by the uniform boundedness principle there are constants $C > 0$ and $M > 0$ such that

$$|\hat{\phi}(\eta)| \leq C(1 + |\eta|)^M.$$

An application of the dominated convergence theorem to the first integral shows that it converges to zero as $k \to \infty$. The second integral converges to zero simply because

$$\sup_{\xi \notin V} (1 + |\xi|)^N |\hat{\phi}_k(\xi)|$$

goes to zero for any N since it is one of the semi-norms that defines the topology on $\mathcal{D}'_\Gamma(K)$. $\qquad \square$

Exercise 3. Where in the above proof do we need that we are dealing with sequences, thus proving only sequential continuity?

Exercise 4. If F is a submersion the pullback can also be constructed by choosing suitable local coordinates. Prove that in this case we have the stronger $\mathrm{WF}(F^*u) = dF^*\mathrm{WF}(u)$.

The Schwartz kernel theorem asserts that the set of continuous linear maps

$$F : \mathcal{D}(\mathcal{U}_1) \to \mathcal{D}'(\mathcal{U}_2)$$

is in one-to-one correspondence with the set of distributions in $\mathcal{D}'(\mathcal{U}_2 \times \mathcal{U}_1)$ in the following sense. Every distributional kernel $K_F \in \mathcal{D}'(\mathcal{U}_2 \times \mathcal{U}_1)$ determines such a map F by

$$F(\psi)(\chi) = K_F(\chi \otimes \psi),$$

where $\chi \otimes \psi$ is the function on $\mathcal{U}_2 \times \mathcal{U}_1$ defined by

$$(\chi \otimes \psi)(x, y) = \chi(x) \cdot \psi(y).$$

Conversely, to every such linear map F there exists a distributional kernel $K_F \in \mathcal{D}'(\mathcal{U}_2 \times \mathcal{U}_1)$ such that $F(\chi)(\psi) = K_F(\psi \otimes \chi)$. One can use this now to define the tensor product of two distributions $u_2 \in \mathcal{D}'(\mathcal{U}_2)$ and $u_1 \in \mathcal{D}'(\mathcal{U}_1)$, namely it is defined as the distributional kernel of the map

$$F : \mathcal{D}(\mathcal{U}_1) \to \mathcal{D}'(\mathcal{U}_2), \quad \phi \mapsto u_1(\phi) \cdot u_2.$$

We write $K_F = u_2 \otimes u_1$.

Exercise 5. Show that

$$\mathrm{WF}(u_2 \otimes u_1) \subset (\mathrm{WF}(u_2) \times \mathrm{WF}(u_1)) \cup ((\mathrm{supp}\, u_2 \times \{0\}) \times \mathrm{WF}(u_1))$$

$$\cup(\mathrm{WF}(u_2) \times (\mathrm{supp}\, u_1 \times \{0\})).$$

Hint: If u_1, u_2 have compact support then the Fourier transform $\hat{u}$ of $u = u_2 \otimes u_1$ is given by $\hat{u}(\xi, \eta) = \hat{u}_2(\xi)\hat{u}_1(\eta)$.

The usual multiplication of functions can be understood as the composition of two maps. If we multiply two functions $f_1, f_2 \in C(\mathcal{U})$ this multiplication can be understood as follows. First we form the tensor product $f_1 \otimes f_2$. This is the function $f_1(x)f_2(y)$ on $\mathcal{U} \times \mathcal{U}$. Then, we restrict the function to the diagonal in $\mathcal{U} \times \mathcal{U}$, that is, we form the pullback of $f_1 \otimes f_2$ under the diagonal map

$$x \mapsto (x, x).$$

Thus, if we ask the question when can we multiply two distributions $u_1, u_2 \in \mathcal{D}'(\mathcal{U})$, we may as well ask, can the distribution $u_2 \otimes u_1 \in \mathcal{D}'(\mathcal{U} \times \mathcal{U})$ be restricted to the diagonal? In order to apply the above criterion we need to calculate the set of normals of the diagonal map. Let us denote the diagonal map by Δ. Then, $d\Delta(\xi) = (\xi, \xi)$ and consequently

$$\langle d\Delta(\xi), (\xi_1, \xi_2) \rangle = \langle \xi, \xi_1 \rangle + \langle \xi, \xi_2 \rangle = \langle \xi, \xi_1 + \xi_2 \rangle.$$

Therefore, $d\Delta^*(\xi_1, \xi_2) = \xi_1 + \xi_2$ and we conclude that the set of normals N_Δ of the diagonal map is given by

$$N_\Delta = \{(x, \xi; x, -\xi) \mid x \in \mathcal{U}, \xi \in \mathbb{R}^n\}.$$

If we apply now our restriction criterion to the distribution $u_2 \otimes u_1$ we get the following.

Corollary 3. *The product $u_1 u_2$ of two distributions $u_1, u_2 \in \mathcal{D}'(\mathcal{U})$ can be defined as the restriction of $u_2 \otimes u_1$ to the diagonal if the following condition is satisfied:*

$$(x, \xi) \in \mathrm{WF}(u_1) \text{ implies } (x, -\xi) \notin \mathrm{WF}(u_2)$$

or in other words $\mathrm{WF}(u_1) \cap -\mathrm{WF}(u_2) = \emptyset$. In this case

$$\mathrm{WF}(u_1 u_2) \subset \mathrm{WF}(u_1) \cup \mathrm{WF}(u_2) \cup (\mathrm{WF}(u_1) + \mathrm{WF}(u_2)),$$

where $\mathrm{WF}(u_1) + \mathrm{WF}(u_2)$ is the set $\{(x, \xi_1 + \xi_2) \mid (x, \xi_1) \in \mathrm{WF}(u_1), (x, \xi_2) \in \mathrm{WF}(u_2)\}$.

In the same way one gets the following nice geometric picture.

Corollary 4. *Let M be a manifold and $N \subset M$ a sub-manifold. Then, a distribution $u \in \mathcal{D}'(M)$ can be restricted to N if the co-normal bundle of N in M does not intersect $\mathrm{WF}(u)$.*

Exercise 6. Show that $\delta_0(|x|^2 - 1)$ is well defined as the pullback of the distribution δ under the map $\mathbb{R}^n \to \mathbb{R}, x \mapsto |x|^2 - 1$. Show that this distribution is the rotation-invariant measure with support on the sphere and use the transformation formula to show that its wavefront set is the normal bundle of the sphere.

4.3.3 Wavefront Sets of the Riesz Distributions

We use the signature convention $(-1, 1, \ldots, 1)$ for Minkowski space and define

$$\gamma(x) = -g(x, x) = x_0^2 - x_1^2 - \cdots - x_{n-1}^2.$$

Then, the wave operator (or d'Alembert operator) $\Box$ is

$$\Box = -\sum_{i,k} g^{ik} \partial_i \partial_k = \partial_0^2 - \partial_1^2 - \cdots - \partial_{n-1}^2.$$

In Chap. 3 it is explained how one constructs a local fundamental solution to a generalized d'Alembert operator from the Riesz distributions $R_\pm(2)$ which were identified as the global retarded and advanced fundamental solutions to the wave equation in Minkowski spacetime. In order to understand the wavefront set of these we would like to calculate the wavefront set of the Riesz distributions $R_\pm(\alpha)$. The Riesz distributions $R_\pm(\alpha)$ were defined as the holomorphic continuations of the regular distributions defined for $\alpha > n$ by

$$R_\pm(\alpha, x) = \begin{cases} C(\alpha, n)\gamma(x)^{\frac{\alpha-n}{2}} & x \in J_\pm(0) \\ 0 & x \notin J_\pm(0) \end{cases},$$

where

$$\gamma(x) = x_0^2 - x_1^2 - \cdots - x_{n-1}^2 \text{ and } C(\alpha, n) := \frac{2^{1-\alpha}\pi^{\frac{2-n}{2}}}{\Gamma(\frac{\alpha}{2})\Gamma(\frac{\alpha-n+2}{2})}.$$

The symbol $J_\pm(0)$ denotes the causal future/past of the point 0, that is the future/past-directed causal vectors in Minkowski space. By the above $R_\pm(\alpha)$ has support in $J_\pm(0)$ and is regular in the interior of $J_\pm(0)$. It is also an immediate consequence of the definition that $R_\pm(\alpha)$ is invariant under the action of the Lorentz group $\mathcal{L}_+^\uparrow$. That the holomorphic continuation actually exists as a distribution in $\mathcal{S}'(\mathbb{R}^n)$ follows from the relation

$$\Box R_\pm(\alpha + 2) = R_\pm(\alpha).$$

Let now $\phi_\alpha(x)$ be the function defined by $\phi_\alpha(x) = x^{\frac{\alpha-n}{2}}$ for $x > 0$ and $\phi_\alpha(x) = 0$ for $x \leq 0$. Then, as long as $\alpha \geq n$ this is a locally integrable function and its wavefront set as a distribution is given by

$$\mathrm{WF}(\phi_\alpha) = \{0\} \times (\mathbb{R}\backslash\{0\}).$$

Indeed, ϕ_α is smooth away from 0 but it is not smooth at zero. So its singular support is $\{0\}$. On the other hand, for any real-valued smooth function $f \in \mathcal{D}(\mathbb{R})$ with $f(0) = 1$ we know that $f\phi_\alpha$ is real valued. Therefore, the Fourier transform satisfies the relation $|\widehat{f\phi_\alpha}(\xi)| = |\widehat{f\phi_\alpha}(-\xi)|$ and we conclude that the wavefront set is symmetric.

In order to understand the singularity structure of the Riesz distribution away from 0 let us restrict $R_\pm(\alpha)$ to the open sets

$$\mathbb{R}_\pm^n = \{(x_0, x_1, \ldots, x_{n-1}) \mid \pm x_0 > 0\}.$$

Now note that the Riesz distribution $R_\pm(\alpha)$ restricted to $\mathbb{R}_\pm^n$ is the pullback of ϕ_α under the submersion $\gamma : \mathbb{R}_\pm^n \to \mathbb{R}$. Therefore, the wavefront set of the restriction of $R_\pm(\alpha)$ to $\mathbb{R}_\pm^n$ is contained in

$$\{(x, \xi) \mid \gamma(x) = 0, \xi \in d\gamma_x^*(\lambda), \lambda \in \mathbb{R}\}.$$

The differential of γ is easily calculated as follows:

$$d\gamma = 2x_0 dx_0 - 2x_1 dx_1 - \cdots - x_{n-1} dx_{n-1},$$

so the adjoint map at x satisfies

$$\langle d\gamma_x^*(\lambda), \eta \rangle = -2\lambda \sum_{i,k} g_{ik} x^i \eta^k.$$

Therefore, using the metric g to identify $\mathbb{R}^n$ with its dual

$$d\gamma_x^*(\lambda) = -2\lambda x.$$

This shows that the wavefront set of the restriction of $R_\pm(\alpha)$ to $\mathbb{R}_\pm^n$ is contained in $\{(x, \xi) \mid \gamma(x) = 0, x = \lambda\xi, \lambda \in \mathbb{R}\}$. Therefore, we have the following.

Proposition 5. *The wavefront set of the Riesz distributions $R_\pm(\alpha)$ is contained in the set*

$$\big(\{0\} \times (\mathbb{R}^n \backslash \{0\})\big) \cup \{(x, \xi) \mid x \in V_\pm, \xi = \lambda x, \lambda \in \mathbb{R}\}.$$

In the above we do not need to assume $\alpha > n$ because $\Box$ is a differential operator and therefore we have $\mathrm{WF}(R_\pm(\alpha)) = \mathrm{WF}(\Box R_\pm(\alpha+2)) \subset \mathrm{WF}(R_\pm(\alpha+2))$. Since the Riesz distributions are invariant under the action of the Lorentz group $\mathcal{L}_+^\uparrow$ so are their wavefront sets. Let us calculate the wavefront set of $R_\pm(2)$, which is particularly interesting since it is the unique global advanced/retarded fundamental solution to the wave equation.

Theorem 11.

$$\mathrm{WF}(R_\pm(2)) = \big(\{0\} \times (\mathbb{R}^n \backslash \{0\})\big) \cup \{(x, \xi) \mid x \in V_\pm, \xi = \lambda x, \xi \neq 0, \lambda \in \mathbb{R}\}.$$

Proof. Since $\Box R_\pm(2) = \delta_0$ and

$$\mathrm{WF}(\delta_0) = \big(\{0\} \times (\mathbb{R}^n \backslash \{0\})\big)$$

we have

$$\big(\{0\} \times (\mathbb{R}^n \backslash \{0\})\big) \subset \mathrm{WF}(R_\pm(2)).$$

It remains to show that also

$$\{(x, \xi) \mid x \in V_\pm, \xi = \lambda x, \xi \neq 0, \lambda \in \mathbb{R}\} \subset \mathrm{WF}(R_\pm(2)).$$

Since $R_\pm(2)$ is real valued its wavefront set is invariant under reflection. Therefore, if (x, x) were not in $\mathrm{WF}(R_\pm(2))$ then x would not be in the singular support of $R_\pm(2)$. However, this can quickly be ruled out by inspecting the explicit
formula. $\square$

4.3.4 Wavefront Set of Other Propagators in Minkowski Space

The Klein–Gordon operator for mass $m > 0$

$$P = \Box + m^2$$

is an example of a generalized d'Alembert operator. Its fundamental solutions can
be explicitly calculated using the Fourier transform. Apart from the retarded and
advanced Green's distributions there are also other Green's distributions and propagators that are used in quantum field theory. These are distinguished by their support
properties in Fourier and in configuration space. In this section I would like to give
a list of propagators and Green's distributions and their wavefront sets. Denote by
G_+ and G_- the retarded and advanced Green's distributions. Since P is translation
invariant the fundamental solutions depend only on the difference of the coordi-
nates. In other words, they are distributions in $\mathcal{D}'(\mathbb{R}^n \times \mathbb{R}^n)$ which are pullbacks of
distributions in $\mathcal{D}'(\mathbb{R}^n)$ under the map

$$\mathbb{R}^n \times \mathbb{R}^n \to \mathbb{R}^n, \quad (x, y) \mapsto x - y.$$

I will denote the corresponding distributions on $\mathbb{R}^n$ by $\tilde{G}_\pm$ to distinguish them nota-
tionally.Thus, $\tilde{G}_\pm$ are the unique distributions satisfying

$$P\tilde{G}_\pm = \delta_0,$$
$$\operatorname{supp} \tilde{G}_\pm = J_\pm(0).$$

The so-called commutator distribution G is then defined as $G_+ - G_-$ and the
corresponding distribution $\tilde{G} \in \mathcal{D}'(\mathbb{R}^n)$ is then a solution to the homogeneous wave
equation. Let $\mu_\pm$ be the following two Lorentz-invariant measures in $\mathbb{R}^n$ with support on the upper/lower mass shell

$$\mu_\pm(\xi) = H(\pm\xi_0)\delta(\gamma(\xi) - m^2).$$

Since $\gamma(\cdot) - m^2$ is a submersion for $\pm\xi_0 > 0$ these measures are indeed well defined
as the pullback of δ under this map. Explicitly, $\mu_\pm$ are given by

$$\int f(x)d\mu_\pm(x) = \int_{\mathbb{R}^{n-1}} \frac{f(\pm E(\underline{x}), \underline{x})}{2E(\underline{x})} d\underline{x},$$

where $E(\underline{x}) = \sqrt{|\underline{x}|^2 + m^2)}$ and $\underline{x}$ denotes the spatial component of x, so that $x = (x_0, \underline{x})$:

$$\tilde{G} = 2\pi i \frac{1}{(2\pi)^{n/2}} (\hat{\mu}_+ - \hat{\mu}_-).$$

From the support properties of $\mathcal{F}(\tilde{G})$ we can see that $\mathrm{WF}(\tilde{G}) \subset \mathbb{R}^n \times (V \backslash \{0\})$, where V is the light-cone. Thus, multiplication with $H(x_0)$ is well defined and we can recover $\tilde{G}^{\pm}$ by

$$\tilde{G}_{\pm}(x) = \pm H(\pm x_0)\tilde{G}(x).$$

The distribution

$$\tilde{\omega}_2 = -2\pi \frac{1}{(2\pi)^{n/2}} \hat{\mu}_+$$

has a Fourier transform with support in the upper mass shell only and can be obtained from $\tilde{G}$ by restricting its Fourier transform to the upper half-plane. Physicists refer to this as the frequency splitting procedure. Another important distribution, the so-called Feynman propagator, arises in perturbation theory and is defined here as

$$G_F(x) = -2\pi i \frac{1}{(2\pi)^{n/2}} (H(x_0)\hat{\mu}_+(x) + H(-x_0)\hat{\mu}_-(x).$$

As a formal expression this can be often found in physics books as

$$\tilde{G}_F(x) = -\frac{i}{(2\pi)^{n-1}} \int_{\mathbb{R}^{n-1}} e^{-i\langle \underline{\xi}, \underline{x} \rangle} \frac{e^{-iE(\underline{\xi})|x_0|}}{2E(\underline{\xi})} d\underline{\xi}.$$

The integral does not converge; one has to understand it as the Fourier transform of a Schwartz distribution as explained earlier. The above-defined distributions have the following meanings in QFT of the free real Klein–Gordon field of mass m:

$$G(f \otimes g) = i[\Phi(f), \Phi(g)]$$
$$G_F(f \otimes g) = i\langle 0|T(\Phi(f)\Phi(g))|\rangle$$
$$\omega_2(f \otimes g) = \langle 0|\Phi(f)\Phi(g)|0\rangle$$

or in a formal sense they appear like this in physics books

$$\tilde{G}(x - y) = i[\Phi(x), \Phi(y)]$$
$$\tilde{G}_F(x - y) = i\langle 0|T(\Phi(x)\Phi(y))|\rangle$$
$$\tilde{\omega}_2(x - y) = \langle 0|\Phi(x)\Phi(y)|0\rangle.$$

These propagators satisfy the following equations:

$$P\hat{G} = 0$$
$$P\hat{\omega}_2 = 0$$
$$P\hat{G}_F = \delta_0$$
$$P\hat{G}_\pm = \delta_0$$

and have wavefront sets

$$\mathrm{WF}(\tilde{G}) = \{(x, \xi) \mid 0 \neq \xi \in V, x = \lambda\xi, \lambda \in \mathbb{R}\}$$
$$\mathrm{WF}(\tilde{\omega}_2) = \{(x, \xi) \mid 0 \neq \xi \in V, x = \lambda\xi, \lambda \in \mathbb{R}, \xi_0 < 0\}$$
$$\mathrm{WF}(\tilde{G}_F) = (\{0\} \times (\mathbb{R}^n \setminus \{0\})) \cup \{(x, \xi) \mid x \in V, \xi = \lambda x, \lambda < 0, \xi \neq 0\}$$
$$\mathrm{WF}(\tilde{G}_\pm) = (\{0\} \times (\mathbb{R}^n \setminus \{0\})) \cup \{(x, \xi) \mid x \in V, \pm x_0 > 0, \xi = \lambda x, \lambda \in \mathbb{R}, \xi \neq 0\}.$$

The above description of the wavefront sets can be obtained in various different ways. For example, as we will see in Sect. 4.5, the advanced and retarded fundamental solutions to any generalized d'Alembert operator in Minkowski space have the same wavefront sets as the Riesz distributions $R_\pm(2)$. Using this, together with the support properties of the Fourier transform, one arrives at the above formulae.

Since the map $(x, y) \mapsto (x - y)$ is a submersion we have

$$\mathrm{WF}(G_i) = \{(x_1, \xi; x_2, -\xi) \mid (x_1 - x_2, \xi) \in \mathrm{WF}(\tilde{G}_i)\}$$

for any of the above propagators G_i.

Exercise 7. Show that the $\tilde{G}$ defined as

$$2\pi\mathrm{i}\frac{1}{(2\pi)^{n/2}}(\hat{\mu}_+ - \hat{\mu}_-)$$

indeed satisfies $P\tilde{G} = 0$ and show that $\tilde{G}_+ = H(x_0)\tilde{G}$ is indeed the retarded Green's distribution.

Exercise 8. Prove that the Feynman propagator is the Fourier transform of the following Schwartz distribution:

$$\lim_{\epsilon \to +0} -\frac{1}{(2\pi)^{n/2}}\frac{1}{\gamma(\xi) - m^2 + \mathrm{i}\epsilon}.$$

Argue why it follows from this that G_F is indeed a Green's distribution, i.e.,

$$P\tilde{G}_F = \delta_0.$$

4.4 Differential Operators, the Wave Equation, and Further Properties of the Wavefront Set

Suppose that

$$P = \sum_{|\alpha| \leq m} a_\alpha(x) \partial^\alpha$$

is a differential operator of order m. Then, the polynomial in ξ

$$p(\xi) = \sum_{|\alpha| \leq m} a_\alpha(x)(i\xi)^\alpha$$

is called the full symbol of P and the principal symbol is the term homogeneous of degree m:

$$\sigma_P(x, \xi) = \sum_{|\alpha|=m} a_\alpha(x)(i\xi)^\alpha.$$

It is not difficult to check that the principal symbol makes sense covariantly as a homogeneous function on the cotangent bundle, whereas the transformation law of the full symbol involves higher order derivatives of the change of charts. The characteristic set char(P) is the set of zeros of σ_P.

Exercise 9. Show that the principal symbol of the wave operator $\Box_g$ in a Lorentzian spacetime is the metric. Show that the characteristic set is the light-cone bundle

$$V = \{(x, \xi) \mid g_x(\xi, \xi) = 0\}.$$

Exercise 10. Suppose that P is a differential operator with constant coefficients and assume that

$$Pu = f,$$

where $u, f \in \mathcal{E}'(\mathcal{U})$. Show that

$$\Sigma(u) \subset \text{char}(P) \cup \Sigma(f).$$

Hint: the symbol $\sigma(P)$ is invertible at points $\xi \notin \text{char}(P)$.

It is actually possible to localize this result and to get the following much stronger statement.

Theorem 12 (Microlocal elliptic regularity). *Suppose that P is a differential operator and $u, f \in \mathcal{D}'(\mathcal{U})$ such that $Pu = f$. Then,*

$$\mathrm{WF}(u) \subset \mathrm{char}(P) \cup \mathrm{WF}(f).$$

The proof becomes tedious if one avoids pseudo-differential operators. As mentioned above one needs to generalize the above exercise to symbols that may depend on x. This can be done within the calculus of pseudo-differential operators, where one allows symbols that are more general than polynomial functions. The main idea is then the same as in the above exercise. One wants to construct an operator Q such that the symbol of QP is one near in a conic neighborhood of $(x, \xi) \notin \mathrm{char}(P)$. The complication comes from the fact that the symbols do not behave multiplicatively if the operators are multiplied. Thus, one cannot simply invert the symbol. Nevertheless, there is an inductive procedure that solves this problem. This procedure can be found in the literature under the name parametrix construction. I will not go into details here but would like to refer the reader to the literature on this subject.

As explained in Chap. 3 a generalized d'Alembert operator on a Lorentzian manifold is an operator which in local coordinates takes the form

$$P = -\sum_{i,k} g^{ik} \frac{\partial^2}{\partial x_i \partial x_k} + Q,$$

where Q is a first-order differential operator. Thus, a generalized d'Alembert operator (or normally hyperbolic operator) is a second-order differential operator whose principal symbol coincides with the metric.

The above theorem shows that any solution $u \in \mathcal{D}'(\mathbb{R}^n)$ to an equation

$$Pu = 0,$$

where P is a generalized *d'Alembert* operator on a Lorentzian manifold has wavefront set contained in the light-cone bundle.

Exercise 11. Show that any distributional solution to the wave equation can be restricted to any spacelike hyper-surface.

There is a second important result which is the celebrated propagation of singularities theorem.

Theorem 13 (Propagation of Singularities). *Let P be a differential operator on a manifold X and suppose that the principal symbol σ_P is real valued. If $u, f \in \mathcal{D}'(X)$ such that*

$$Pu = f,$$

then $\mathrm{WF}(u) \backslash \mathrm{WF}(f)$ *is invariant under the local flow on* $T^*X \backslash \mathrm{WF}(f)$ *generated by the Hamiltonian vector field of the function* σ_P.

This theorem can be proved using Egorov's theorem in the theory of pseudo-differential operators. I will not prove it here but I would like to point out here that this establishes a link between the classical world and the quantum world at high energies. If quantum dynamics is described by an operator P then the high-energy classical limit is described by the Hamiltonian flow of its principal symbol.

Let us specify what this means if P is a generalized d'Alembert operator on a Lorentzian manifold X. Then, $\sigma_P(\xi) = g(\xi, \xi)$ and the Hamiltonian flow is, in local coordinates (x_i, ξ_i) of T^*X, generated by the vector field

$$\sum_k \frac{\partial g_x(\xi, \xi)}{\partial \xi_k} \frac{\partial}{\partial x_k} - \frac{\partial g_x(\xi, \xi)}{\partial x_k} \frac{\partial}{\partial \xi_k} = \sum_k 2g^{ik}\xi_i \frac{\partial}{\partial x_k} - \frac{\partial g^{mn}}{\partial x_k}\xi_m\xi_n \frac{\partial}{\partial \xi_k}.$$

Therefore, $(x(t), \xi(t))$ is a solution to the flow equation if it satisfies the equation

$$\frac{d}{dt}\begin{pmatrix} x_k \\ \xi_k \end{pmatrix} = \begin{pmatrix} \sum_i 2g^{ik}\xi_i, \\ \sum_{m,n} -\frac{\partial g^{mn}}{\partial x_k}\xi_m\xi_n \end{pmatrix}.$$

Using $\frac{dg_{nm}}{dt} = \sum_\alpha \frac{\partial g_{mn}}{\partial x_\alpha} \frac{dx_\alpha}{dt}$ one can see quickly that a solution curve $x(t)$ satisfies the equation for a geodesic

$$\frac{d^2 x_k}{dt^2} + \sum_{m,n} \Gamma_{mn}^k \frac{dx_m}{dt} \frac{dx_n}{dt} = 0,$$

where

$$\Gamma_{mn}^k = \sum_l \frac{1}{2} g^{kl} \left(\frac{\partial g_{ml}}{\partial x_n} + \frac{\partial g_{nl}}{\partial x_m} - \frac{\partial g_{nm}}{\partial x_l} \right)$$

are the Christoffel symbols. This shows that the above Hamiltonian flow is the geodesic flow and the Hamiltonian vector field is the geodesic spray. Thus, we have the following.

Corollary 5. *Suppose that X is a Lorentzian manifold and $u, f \in \mathcal{D}'(X)$ such that*

$$Pu = f,$$

where P is a generalized d'Alembert operator. Suppose furthermore that $(x_0, \xi_0) \in$ WF(u) and suppose that there is a geodesic $\gamma : [0, T] \to X$ such that

- $\gamma(0) = x_0$,
- $\dot{\gamma}(0) = \xi_0$,
- $(\gamma(t), \dot{\gamma}(t)) \notin$ WF(f) *for $t \in [0, T]$.*

Then, $(\gamma(t), \dot{\gamma}(t)) \in$ WF(u) for all $t \in [0, T]$.

Or in simple terms singularities of solutions to the wave equation travel along lightlike geodesics. This provides the reason why light travels on lightlike geodesics. Note that in curved spacetimes Huygen's principle does in general not hold: there, in general, are residual waves from back-scattering of the light at the non-constant curvature so that the smooth part of a solution may travel at a lower speed. The above theorem guarantees, however, that singularities travel at the speed of light. It also shows that in the short wave limit wave optics becomes geometrical optics. I would like to point out here that there are proofs of this which rely on constructions from geometrical optics (see [5] for an excellent introduction on the Fourier integral operator approach).

4.5 Wavefront Set of Propagators in Curved Spacetimes

Now let us turn to the local retarded and advanced Green's distributions $G_\pm$ for a generalized d'Alembert operator P in a Lorentzian spacetime, i.e., the metric g has signature $(-1, 1, \ldots, 1)$ and we define $\gamma_x(\xi) = -g(\xi, \xi)$. Suppose that Ω' is a geodesically convex open set. This means that the inverse of the exponential map

$$\exp^{-1} : \Omega' \times \Omega' \to T\Omega'$$

is a diffeomorphism onto a neighborhood of the zero section in $T\Omega'$. We may choose a time-oriented orthonormal frame to identify

$$T\Omega' \cong \Omega' \times \mathbb{R}^n$$

and denote the projection onto the second factor by p. Then, p is a submersion and we may pullback the Riesz distribution $R_\pm(\alpha)$ to a distribution on $T\Omega'$. Since the Riesz distribution is Lorentz invariant the obtained distribution on $T\Omega'$ does not depend on the choice of the frame. The pullback of this distribution under $\exp^{-1}$ is then well defined and called Riesz distribution $R_\pm^{\Omega'}(\alpha)$ on $\Omega' \times \Omega'$. So $R_\pm^{\Omega'}(\alpha)$ is the pullback of $R_\pm(\alpha)$ under the map

$$F : \Omega' \times \Omega' \longrightarrow T\Omega' \longrightarrow \Omega' \times \mathbb{R}^n \longrightarrow \mathbb{R}^n.$$

Alternatively, one may define $R_\pm^{\Omega'}$ on $\Omega' \times \Omega'$ also as the analytic continuation in the sense of distributions of the functions

$$R_\pm^{\Omega'}(\alpha; x, y) = \begin{cases} C(\alpha, n)\Gamma(x, y)^{\frac{\alpha-n}{2}} & y \in J_\pm(x) \\ 0 & y \notin J_\pm(x) \end{cases}.$$

As before $J_\pm(x)$ is the causal future/past of the point x. One can see this, for example, by defining them directly as a pullback of the Riesz distributions. In Chap. 3 we learned that if $\Omega \subset \Omega'$ is an open relatively compact causal subset of Ω' with $\overline{\Omega} \subset \Omega'$ then there are unique local retarded and advanced Green's distributions $G_\pm^\Omega$ to P on $\Omega \times \Omega$. They can be described asymptotically by the so-called Hadamard series, which expresses $G_\pm^\Omega$ as an asymptotic sum over Riesz distributions. More precisely, for $N \geq n/2$ and for $k \in \mathbb{N}$ one considers the partial sums

$$\mathcal{R}_{\pm}^{N+k} := \sum_{j=0}^{N+k-1} V^j \cdot R_{\pm}^{\Omega'}(2j+2),$$

where $(V^j)_{j\geq 0} \in C^\infty(\Omega' \times \Omega')$ is the sequence of Hadamard coefficients for P on Ω'. Then, Proposition 5 on page 72 states that for every $k \in \mathbb{N}$ the map

$$(x, y) \mapsto (G_{\pm}^{\Omega}(x) - \mathcal{R}_{\mp}^{N+k}(x))(y)$$

is a C^k-function on $\overline{\Omega} \times \overline{\Omega}$.

Since the Fourier transform of compactly supported C^k function can be bounded by $C(1 + |\xi|)^{-k}$ this shows that the wavefront set of $G_{\pm}^{\Omega}$ cannot be larger than the union of all the wavefront sets of all $R_{\mp}^{\Omega}(2j+2)$.

Lemma 5. *If* $(x, \xi; x, \eta) \in \mathrm{WF}(R_{\pm}^{\Omega'}(\alpha))$ *then* $\xi = -\eta$. *Similarly, if* $x \in \Omega$ *and* $(x, \xi; x, \eta) \in \mathrm{WF}(G_{\pm}^{\Omega})$ *then* $\xi = -\eta$.

Proof. By the above the second statement follows from the first. For the first statement we recall the definition of $R_{\pm}^{\Omega'}(\alpha)$ as the pullback of $R_{\pm}(\alpha)$ under the map

$$F : \Omega' \times \Omega' \longrightarrow T\Omega' \longrightarrow \Omega' \times \mathbb{R}^n \longrightarrow \mathbb{R}^n.$$

The differential of F at the diagonal is easily calculated and equals

$$dF(x, x)(\xi_1, \xi_2) = \xi_2 - \xi_1.$$

Thus, the adjoint is $dF^*(\xi) = (-\xi, \xi)$. The rest follows from $\mathrm{WF}(R_{\pm}^{\Omega'}(\alpha)) \subset dF^*\mathrm{WF}(R_{\pm}(\alpha))$. $\qquad\square$

Lemma 6. *Let* $G_{\pm}^{\Omega}$ *be the local retarded and advanced Green's distributions for a generalized d'Alembert operator* P *on* $\Omega \times \Omega$. *Then,* $(x, \xi; x, \eta) \in \mathrm{WF}(G_{\pm}^{\Omega})$ *if and only if* $\xi = -\eta \neq 0$.

Proof. By definition of Green's distributions we have $(P \otimes \mathrm{id})G_{\pm}^{\Omega} = \delta_\Delta$ where δ_Δ is the Dirac distribution supported on the diagonal. This means that $\delta_\Delta(f) = \int_M f(x, x)dx$. As we already saw in local coordinates δ_Δ has wavefront set equal to the normal bundle of the diagonal, i.e.,

$$\mathrm{WF}(\delta_\Delta) = \{(x, -\xi; x, \xi) \mid \xi \in \mathbb{R}^n \setminus \{0\}\}$$

and by the above $\mathrm{WF}(\delta_\Delta)$ is contained in the wavefront set of $G_{\pm}^{\Omega}$. $\qquad\square$

Lemma 7. *For* $(x_1, x_2) \in \Omega \times \Omega$ *with* $x_1 \neq x_2$ *a point* $(x_1, \xi_1, x_2, -\xi_2)$ *is in* $\mathrm{WF}(G_{\pm}^{\Omega})$ *if and only if* $x_2 \subset J_{\pm}(x_1)$ *and there is a lightlike geodesic in* Ω' *through* x_1 *and* x_2 *with tangent* ξ_1 *at* x_1 *and* ξ_2 *at* x_2, *i.e.,* (x_1, ξ_1) *and* (x_2, ξ_2) *are in the same orbit of the geodesic flow on* $T\Omega'$.

Proof. Let us assume $x_2 \in J_\pm(x_1)$ since otherwise we are outside the support of $G_\pm^\Omega$. Since $x_1 \neq x_2$ the map $(x, y) \to \Gamma(x, y)$ is a submersion near (x_1, x_2) and we can find local coordinates $t_1, t_2, \ldots, t_{2n}$ such that $t_2, \ldots, t_{2n}$ parametrize the set where $\Gamma(x, y) = 0$ and $\Gamma(\mathbf{t}) = t_1$. Then, up to a non-vanishing smooth factor $R_\pm^\Omega(\alpha)$ is given by the analytic continuation of $C(\alpha, n)\mathrm{sign}(t)t_1^{\frac{\alpha-n}{2}}$. For $\alpha = 2$ this is singular at the hyper-surface $t_1 = 0$. For $\alpha > 2$ there might as well be a singularity at $t_1 = 0$, but it is of strictly smaller order. If we understand $C(\alpha, n)\mathrm{sign}(t_1)t_1^{\frac{\alpha-n}{2}}$ as distributions in one variable t_1 we have that if

$$u(t) - \sum_{j=0}^{N} C(2j+2, n)g_j(t_1)\mathrm{sign}(t_1)t_1^{\frac{2j+2-n}{2}} \in C^0$$

for N large and some smooth functions g_j with $g_0(0) \neq 0$ then

$$\mathrm{WF}(u) = \mathrm{WF}(C(2, n)g_0(t_t)\mathrm{sign}(t_1)t_1^{\frac{2-n}{2}}) = \{0\} \times (\mathbb{R}\backslash\{0\}),$$

as only the first term determines the behavior of the Fourier transform at infinity. Applied to our situation, where

$$\mathcal{R}_\mp^{N+k} := \sum_{j=0}^{N+k-1} V^j \cdot R_\mp^{\Omega'}(2j+2)$$

is on $\Omega \times \Omega$ an approximation to $G_\pm^\Omega$ modulo C^k-functions, we conclude that if $x_1 \neq x_2$ then $(x_1, \xi_1; x_2, \xi_2) \in \mathrm{WF}(G_\pm^\Omega)$ if and only if $(x_1, \xi_1; x_2, \xi_2) \in \mathrm{WF}(R_\mp^{\Omega'}(2))$. The above local calculation means that if $x_1 \neq x_2$ and $\Gamma(x_1, x_2) = 0$ and $x_2 \subset J_\pm(x_1)$, $(x_1, \xi_1; x_2, \xi_2) \in \mathrm{WF}(R_\pm^{\Omega'}(2))$ if $(\xi_1, \xi_2) = d\Gamma^*(x_1, x_2)(\lambda)$ for some $\lambda \neq 0$. It remains to determine the image of $d\Gamma^*(x_1, x_2)$. By the Gauss lemma

$$d\Gamma(x_1, x_2)(\xi_1, \xi_2) = -2\langle d\exp(\eta_2)(\eta_2), \xi_1 \rangle - 2\langle d\exp(\eta_1)(\eta_1), \xi_2 \rangle,$$

where $\eta_2 = \exp_{x_2}^{-1}(x_1)$ and $\eta_1 = \exp_{x_1}^{-1}(x_2)$. Since η_1 and $-\eta_2$ are in the same orbit of the geodesic flow it follows that $(\xi_1, \xi_2) \in d\Gamma^*(x_1, x_2)(\mathbb{R}\backslash\{0\})$ if and only if (x_1, ξ_1) and $(x_2, -\xi_2)$ are in the same orbit of the geodesic flow. $\square$

The above gives us a complete description of the wavefront set of the local Green's distributions in Ω.

Theorem 14. *Let P be a generalized d'Alembert operator on a Lorentzian manifold and let $G_\pm^\Omega \in \mathcal{D}'(\Omega \times \Omega)$ the local Green's distributions as above. Then,*

$$\mathrm{WF}(G_\pm^\Omega) = \{(x, \xi; x, -\xi) \mid (x, \xi) \in T^*M\backslash 0\}\cup$$

$$\cup\{(x_1, -\xi_1; x_2, \xi_2) \mid (x_1, \xi_1) \sim (x_2, \xi_2), \ x_2 \in J_\pm(x_1), \gamma_{x_1}(\xi_1) = 0, \xi_1 \neq 0\},$$

where $(x_1, \xi_1) \sim (x_2, \xi_2)$ *means that* (x_1, ξ_1) *and* (x_2, ξ_2) *are in the same orbit of the geodesic flow on* $T\Omega'$.

This can now be used to derive propagation of singularities for the wave equation as well as microlocal elliptic regularity. I will not follow this path here but would rather like to assume these theorems. In a certain sense the special form of the wave-front set of a Green's distribution is equivalent to propagation of singularities and microlocal elliptic regularity. In fact, on a globally hyperbolic spacetime the singularity structure of $G_\pm$ near the diagonal together with the propagation of singularities theorem already determines the wavefront set as the proof of the following theorem will show.

Theorem 15. *Let* P *be a generalized d'Alembert operator on a globally hyperbolic spacetime* (M, g) *and denote by* $G_\pm \in \mathcal{D}'(M \times M)$ *the unique global retarded and advanced Green's distributions. Then,*

$$\mathrm{WF}(G_\pm) = \{(x, \xi; x, -\xi) \mid (x, \xi) \in T^*M \backslash 0\} \cup$$
$$\cup \{(x_1, \xi_1; x_2, -\xi_2) \mid (x_1, \xi_1) \sim (x_2, \xi_2), \ \gamma_{x_1}(\xi_1) = 0, \ x_2 \in J_\pm(x_1), \xi_1 \neq 0\},$$

where $(x_1, \xi_1) \sim (x_2, \xi_2)$ *means that* (x_1, ξ_1) *and* (x_2, ξ_2) *are in the same orbit of the geodesic flow, i.e., there is a geodesic through both* x_1 *and* x_2 *with tangent* ξ_1 *and ending at* x_2 *with tangent* ξ_2.

Proof. Since each point x has a causal neighborhood which is relatively compact on a convex open neighborhood, Theorem 14 determines the wavefront set near the diagonal. In particular, on the diagonal the wavefront set is the same as the wavefront set of the local Green's distributions. Since we already determined the wavefront set on the diagonal it remains to look at the points away from the diagonal. Let us do the proof for G_+ only since the proof for G_- works with the obvious modifications. Away from the diagonal both $(\mathrm{id} \otimes P^t)G_\pm$ and $(P \otimes \mathrm{id})G_\pm = 0$ vanish. Thus, $(x_1, \xi_1; x_2, \xi_2) \in \mathrm{WF}(G_+)$ with $x_1 \neq x_2$ implies that $g(\xi_1, \xi_1) = g(\xi_2, \xi_2) = 0$. We can also apply the propagation of singularities theorem in both variables away from the diagonal. Let (x_2, ξ_2) be any point in T^*M and let $(x_1, \xi_1) \in T^*M$ such that $x_1 \neq x_2, g(\xi_1, \xi_1) = g(\xi_2, \xi_2) = 0$, and $\xi_1 \neq 0$. Since M is globally hyperbolic there exists a Cauchy surface S which contains the point x_2. Let $s(t)$ be an in-extendible lightlike geodesic through x_1 with tangent ξ_1. Since $s(t)$ is a causal curve it will intersect S precisely in one point x' with tangent ξ'. Now there are several possibilities:

- $x' \neq x_2$. Since the support of G_+ does not contain the points (x', x_2) with $x_2 \neq x'$ the point (x', ξ', x_2, ξ_2) is not in the wavefront set of G_+. By the propagation of singularity theorem then also $((x_1, \xi_1; x_2, \xi_2))$ is not in the wavefront set of G_+.
- $x' = x_2$ and $x_1 \notin J_+(x_2)$: The point (x_1, x_2) is not in the support of G_+ and therefore, $((x_1, \xi_1; x_2, \xi_2))$ is not in the wavefront set of G_+.
- $x' = x_2, x_1 \in J_+(x_2)$, and $\xi_2 \neq -\xi'$: For $x_1 \in J_+(x_2)$ the point (x_1, x_2) is not in the support of G_- and therefore near this point we can as well look at the structure of the commutator distribution $G = G_+ - G_-$ which solves the wave equation everywhere. Thus, if $((x_1, \xi_1; x_2, \xi_2))$ were in the wavefront set of G_+ then, by

propagation of singularities, also (x', ξ', x_2, ξ_2) would be in the wavefront set of WF(G). But $(x, x, \xi, \eta) \in$ WF(G) implies by Lemma 6 that $\xi = -\eta$. Therefore, if $\xi_1 \neq -\xi'$ then $((x_1, \xi_1; x_2, \xi_2))$ is not in WF(G_+).

- $x' = x_2$, $x_1 \in J_+(x_2)$ and $\xi_1 = -\xi'$. We want to show that in this case $(x_1, \xi_1; x_2, \xi_2) \in$ WF(G_+). The propagation of singularities theorem will imply this if we can show that $(x_3, -\xi_3; x_2, \xi_2) \in$ WF(G_+) for some $(x_3, \xi_3) \sim (x_2, \xi_2)$ with $x_3 \subset J_+(x_2)$ and x_3 on the geodesic connecting x_1 and x_2. We may choose x_3 close enough to x_2 such that they both are contained in causal set Ω which is relatively compact in a convex set Ω'. By uniqueness of Green's distributions, G_+ restricted to $\Omega \times \Omega$ coincides with the local fundamental solution $G_\pm^\Omega$. By Theorem 14 $(x_3, -\xi_3; x_2, \xi_2) \in$ WF(G_+).

It remains to prove that $(x_1, 0; x_2, \xi_2)$ is not in WF(G_+). This can be shown using the same argument: if it were in WF(G_+) then $\xi_2 \neq 0$ and $(x_1, \xi_1; x_3, \xi_3) \in$ WF(G_+) if (x_3, ξ_3) is in the same orbit of the geodesic flow as (x_2, ξ_2). Choose a Cauchy surface S through x_1 and (x_3, ξ_3) on this Cauchy surface. Then, $(x_2, 0, x_3, \xi_3) \in$ WF(G_+). In case $x_2 = x_3$ this contradicts Lemma 6. In case $x_2 \neq x_3$ this contradicts $(x_2, x_3) \notin$ supp (G_+). Therefore, $(x_1, 0; x_2, \xi_2) \notin$ WF(G_+). $\qquad \square$

Analogously we obtain for the wavefront set of G.

Theorem 16. *Let $P, M, G_\pm$ as above and let $G = G_+ - G_-$ be the commutator distribution. Then,*

$$\text{WF}(G) = \{(x_1, -\xi_1; x_2, \xi_2) \mid (x_1, \xi_1) \sim (x_2, \xi_2), \xi_1 \neq 0\}.$$

So the wavefront sets agree with the obvious generalizations from the Minkowski case. The states for a quantum field that are physically reasonable are believed to have the analogous wavefront sets as their Minkowski counterparts. This motivates the following definition.

Definition 10. *A bi-solution ω_2 to P is called of Hadamard form if*

$$\text{WF}(\omega_2) = \{(x_1, -\xi_1; x_2, \xi_2) \mid (x_1, \xi_1) \sim (x_2, \xi_2), (\xi_1)_0 > 0\}.$$

For such a distribution to be a state it needs to satisfy the additional property $\omega_2(\overline{f} \otimes f) \geq 0$. The existence of such a Hadamard state can be established for the Klein–Gordon field on any globally hyperbolic spacetime. Specification of such a state is equivalent to the splitting procedure in Sect. 4.3.4. Note, however, that Hadamard states are not unique since a restriction on the wavefront set is only an asymptotic condition on the energy.

Exercise 12. Suppose that we are given a Hadamard state ω_2 of the Klein–Gordon field on a globally hyperbolic spacetime. Define a Feynman propagator, using only ω_2, and compute its wavefront set.

References

1. Hörmander, L.: The analysis of linear partial differential operators, I–III. Springer-Verlag, Berlin (2007)
2. Hörmander, L.: Fourier integral operators I. Acta Math. **127**, 79 (1971)
3. Duistermaat, J., Hörmander, L.: Fourier Integral Operators II. Acta Math. **128**, 183 (1972)
4. Radzikowski, M.: The Hadamard Condition and Kay's conjecture in (axiomatic) Quantum Field Theory on curved spacetimes, Thesis. Princeton University, Princeton (1992)
5. Grigis, A., Sjöstrand, J.: Microlocal Analysis for Differential Operators, London Mathematical Society Lecture Note Series 196, Cambridge University Press, Cambridge (1994)

Chapter 5
Quantum Field Theory on Curved Backgrounds

Romeo Brunetti and Klaus Fredenhagen

5.1 Introduction

Quantum field theory is an extremely successful piece of theoretical physics. Based
on few general principles, it describes with an incredibly good precision large parts
of particle physics. But also in other fields, in particular in solid state physics,
it yields important applications. At present, the only problem which seems to go
beyond the general framework of quantum field theory is the incorporation of gravity. Quantum field theory on curved backgrounds aims at a step toward solving this
problem by neglecting the back reaction of the quantum fields on the spacetime
metric.

Quantum field theory has a rich and rather complex structure. It appears in different versions that are known to be essentially equivalent. Unfortunately, large parts
of the theory are available only at the level of formal perturbation theory, and a
comparison of the theory with experiments requires a truncation of the series which
is done with a certain arbitrariness.

Due to its rich structure, quantum field theory is intimately related to various
fields of mathematics and has often challenged the developments of new mathematical concepts.

In this chapter we will give an introduction to quantum field theory in a formulation which admits a construction on generic spacetimes. Such a construction
is possible in the so-called algebraic approach to quantum field theory [1, 2]. The
more standard formulation as one may find it in typical text books (see, e.g., [3])
relies heavily on concepts like vacuum, particles, energy, and makes strong use of
the connection to statistical mechanics via the so-called Wick rotation. But these
concepts lose their meaning on generic Lorentzian spacetimes and are therefore

R. Brunetti (✉)
Dipartimento di Matematica, Università di Trento, Via Sommarive 14, I-38050 Povo (TN)
e-mail: brunetti@science.unitn.it; romeo.brunetti@gmail.com

K. Fredenhagen (✉)
II. Institut für Theoretische Physik, Universität Hamburg, Luruper Chaussee 149,
D-22761 Hamburg, Germany
e-mail: klaus.fredenhagen@desy.de

Brunetti, R., Fredenhagen, K.: *Quantum Field Theory on Curved Backgrounds*. Lect. Notes
Phys. **786**, 129–155 (2009)
DOI 10.1007/978-3-642-02780-2_5   © Springer-Verlag Berlin Heidelberg 2009

restricted to a few examples with high symmetry. It was a major progress of recent years that local versions of most of these concepts have been found. Their formulation requires the algebraic framework of quantum physics and, on the more technical side, the replacement of momentum space techniques by techniques from microlocal analysis.

The plan of the chapter is as follows. After a general discussion of fundamental physical concepts like states, observables, and subsystems we will describe a general framework that can be used to define both classical and quantum field theories. It is based on the locally covariant approach to quantum field theory [4] which uses the language of categories to incorporate the principle of general covariance.

The first example of the general framework is the canonical formalism of classical field theory based on the so-called Peierls bracket by which the algebra of functionals of classical field configurations is endowed with a Poisson structure.

We then present as a simple example in quantum field theory the free scalar quantum field.

A less simple example is the algebra of Wick polynomials of the free field. Here, for the first time, techniques from microlocal analysis enter. The construction relies on a groundbreaking observation of Radzikowski [5]. Radzikowski found that the so-called Hadamard condition on the two-point correlation function is equivalent to a positivity condition on the wave front set, whose range of application was extended and named "microlocal spectrum condition" few years later [6]. This insight not only, for the first time, permitted the construction of nonlinear fields on generic spacetimes but also paved the way for a purely algebraic construction, which before was also unknown on Minkowski space.

Based on these results, one now can construct also interacting quantum field theories in the sense of formal power series. The construction can be reduced to the definition of time-ordered products of prospective Lagrangians. By the principle of causality, the time-ordered products of n factors are determined by products (in the sense of the algebra of Wick polynomials) of time-ordered products of less than n factors outside of the thin diagonal $\Delta_n \subset \mathcal{M}^n$ (considered as algebra-valued distributions). The removal of ultraviolet divergences amounts in this framework to the extension of distributions on $\mathcal{M}^n \setminus \Delta_n$ to $\mathcal{M}^n$. The possible extensions can be discussed in terms of the so-called microlocal scaling degree which measures the singularity of the distribution transversal to the submanifold Δ_n.

5.2 Systems and Subsystems

5.2.1 Observables and States

Experiments on a physical system may be schematically described as maps

$$\text{experiment} : (\text{state}, \text{observable}) \mapsto \text{result}. \tag{5.1}$$

Here a state is understood as a prescription for the preparation of the system, and the observable is an operation on the prepared system which yields a definite result. In classical physics, one assumes that an optimally prepared system (pure state) yields for a given (ideal) observable always the same result (which may be recorded as a real number). Thus observables can be identified with real-valued functions on the set of pure states. The set of observables so gets the structure of an associative, commutative algebra over $\mathbb{R}$, and the pure states are reobtained as characters of the algebra, i.e., homomorphisms into $\mathbb{R}$.

In classical statistical mechanics one considers also incomplete preparation prescriptions, e.g., one puts a number of particles into a box with a definite total energy, but without fixing positions and momenta of the individual particles. Such a state corresponds to a probability measure μ on the set of pure states, or, equivalently, to a linear functional on the algebra of observables which is positive on positive functions and assumes the value 1 on the unit observable. For the observable f the state yields the probability distribution

$$(\mu, f) \mapsto f_* \mu , \quad f_* \mu(I) = \mu(f^{-1}(I)) \tag{5.2}$$

on $\mathbb{R}$. Pure states are the Dirac measures.

In quantum mechanics, the measurement results fluctuate even in optimally prepared states. Pure states are represented by one-dimensional subspaces $\mathfrak{L}$ of some complex Hilbert space, and observables are identified with self-adjoint operators A. The probability distribution of measured values is given by

$$\mu_{A,\mathfrak{L}}(I) = (\Psi, E_A(I)\Psi), \tag{5.3}$$

where Ψ is any unit vector in $\mathfrak{L}$ and $E_A(I)$ is the spectral projection of A corresponding to the interval I.

In quantum statistics, one admits a larger class of states, corresponding to incomplete preparation, which can be described by a density matrix, i.e., a positive trace class operator ρ with trace 1; the probability distribution is given by

$$\mu_{A,\rho}(I) = \mathrm{Tr} \rho E_A(I), \tag{5.4}$$

where the pure states correspond to the rank 1 density matrices.

In spite of the apparently rather different structures one can arrive at a unified description. The set of observables is a real vector space with two products:

1. a commutative, but in general nonassociative product (the Jordan product),

$$A \circ B = \frac{1}{4} \left((A + B)^2 - (A - B)^2 \right), \tag{5.5}$$

   arising from the freedom of relabeling measurement results;
2. an antisymmetric product

$$\{A, B\}, \tag{5.6}$$

which is known as the Poisson bracket in classical mechanics and is given by $\frac{i}{\hbar}$ times the commutator $[\cdot, \cdot]$ in quantum mechanics. This product originates from the fact that every observable H can induce a transformation of the system by Hamilton's (or Heisenberg's) equation

$$\frac{d}{dt}A(t) = \{H, A(t)\}. \tag{5.7}$$

The two products satisfy the following conditions:

1. $A \mapsto \{B, A\}$ is a derivation with respect to both products.
2. The associators of both products are related by

$$(A \circ B) \circ C - A \circ (B \circ C) = \frac{\hbar^2}{4}\left(\{\{A, B\}, C\} - \{A, \{B, C\}\}\right). \tag{5.8}$$

While the first condition is motivated by the interpretation of Hamilton's equation as an infinitesimal symmetry, there seems to be no physical motivation for the second condition. But mathematically, it has a strong impact: in classical physics $\hbar = 0$, hence the Jordan product is associative; in quantum physics, the condition implies that

$$AB := A \circ B + \frac{\hbar}{2i}\{A, B\} \tag{5.9}$$

is an associative product on the complexification $\mathfrak{A} = \mathfrak{A}_\mathbb{R} \otimes \mathbb{C}$, where the information on the real subspace is encoded in the $\star$-operation

$$(A \otimes z)^* = A \otimes \bar{z}. \tag{5.10}$$

States are defined as linear functionals on the algebra which assume positive values on positive observables and are 1 on the unit observable. A priori, in the case $\hbar \neq 0$ the positivity condition on the subspace $\mathfrak{A}_\mathbb{R}$ of self-adjoint elements could be weaker than the positivity requirement on the complexification $\mathfrak{A}$. Namely, on the real subspace we call positive every square of a self-adjoint element, whereas on the full algebra positive elements are absolute squares of the form

$$(A - iB)(A + iB) = A^2 + B^2 + \hbar\{A, B\}, \quad A, B \text{ self-adjoint}. \tag{5.11}$$

But under suitable completeness assumptions, in particular when $\mathfrak{A}$ is a C^*-algebra, operators as above admit a self-adjoint square root; thus the positivity conditions coincide in these cases. If one is in a more general situation, one has to require that states satisfy the stronger positivity condition, in order to ensure the existence of the GNS representation.

5.2.2 Subsystems

A system may be identified with a unital C^*-algebra $\mathfrak{A}$. Subsystems correspond to sub-C^*-algebras $\mathfrak{B}$ with the same unit. A state of a system then induces a state on the subsystem by restricting the linear functional ω on $\mathfrak{A}$ to the subalgebra $\mathfrak{B}$. The induced state may be mixed even if the original state was pure, see Remark 13 on page 23.

One may also ask whether every state on the subalgebra $\mathfrak{B}$ arises as a restriction of a state on $\mathfrak{A}$. This is actually true, namely let ω be a state on $\mathfrak{B}$. According to the Hahn–Banach theorem, ω has an extension to a linear functional $\tilde{\omega}$ on $\mathfrak{A}$ with $\|\tilde{\omega}\| = \|\omega\|$. But $\tilde{\omega}(1) = \omega(1) = \|\omega\| = 1$, hence $\tilde{\omega}$ is a state.

Two subsystems $\mathfrak{B}_1$ and $\mathfrak{B}_2$ may be called independent whenever the algebras $\mathfrak{B}_1$ and $\mathfrak{B}_2$ commute and

$$B_1 \otimes B_2 \mapsto B_1 B_2 \tag{5.12}$$

defines an isomorphism from the tensor product $\mathfrak{B}_1 \otimes \mathfrak{B}_2$ to the algebra generated by $\mathfrak{B}_1$ and $\mathfrak{B}_2$.

Given states ω_i on $\mathfrak{B}_i$, $i = 1, 2$, one may define a product state on $\mathfrak{B}_1 \otimes \mathfrak{B}_2$ by

$$(\omega_1 \otimes \omega_2)(B_1 \otimes B_2) = \omega_1(B_1)\omega(B_2), \tag{5.13}$$

see Section 5.5 on page 23 for a thorough discussion. Convex combinations of product states are called separable. As was first observed by Bell, there exist nonseparable states if both algebras contain subalgebras isomorphic to $M_2(\mathbb{C})$. This is the famous phenomenon of entanglement which shows that states in quantum physics may exhibit correlations between independent systems which cannot be described in terms of states of the individual systems. This is the reason, why the notion of locality is much more evident on the level of observables than on the level of states.

5.2.3 Algebras of Unbounded Operators

In applications often the algebra of observables cannot be equipped with a norm. The CCR algebra is a prominent example. In these cases one usually still has a unital $*$-algebra, and states can be defined as positive normalized functionals. The GNS construction remains possible, but does not lead to a representation by bounded Hilbert space operators. In particular it is not guaranteed that self-adjoint elements of the algebra are represented by self-adjoint Hilbert space operators. There is no general theory available which yields a satisfactory physical interpretation in this situation. One therefore should understand it as an intermediary step toward a formulation in terms of C^*-algebras.

5.3 Locally Covariant Theories

5.3.1 Axioms of Locally Covariant Theories

Before constructing examples of classical and quantum field theories we want to describe the minimal requirements that such theories should satisfy the following [4]:

1. To each globally hyperbolic time-oriented spacetime $\mathcal{M}$ we associate a unital $*$-algebra $\mathfrak{A}(\mathcal{M})$.
2. Let $\chi : \mathcal{M} \to \mathcal{N}$ be an isometric embedding which preserves causal relations in the sense that whenever $\chi(x) \in J_+^{\mathcal{N}}(\chi(y))$ for some points $x, y \in \mathcal{M}$ then $x \in J_+^{\mathcal{M}}(y)$. Then there is an injective homomorphism

$$\alpha_\chi : \mathfrak{A}(\mathcal{M}) \to \mathfrak{A}(\mathcal{N}). \tag{5.14}$$

3. Let $\chi : \mathcal{M} \to \mathcal{N}$ and $\chi' : \mathcal{N} \to \mathcal{L}$ be causality-preserving isometric embeddings. Then

$$\alpha_{\chi' \circ \chi} = \alpha_{\chi'} \alpha_\chi. \tag{5.15}$$

These axioms characterize a quantum field theory as a covariant functor $\mathfrak{A}$ from the category $\mathfrak{Man}$ of globally hyperbolic time-oriented Lorentzian manifolds with isometric causality-preserving mappings as morphisms to the category of unital $*$-algebras $\mathfrak{Alg}$ with injective homomorphisms as morphisms, where $\mathfrak{A}$ acts on morphisms by $\mathfrak{A}\chi = \alpha_\chi$.

In addition we require

4. Let $\chi_i : \mathcal{M}_i \to \mathcal{N}, i = 1, 2$, be morphisms with causally disjoint images. Then the images of $\mathfrak{A}(\mathcal{M}_1)$ and $\mathfrak{A}(\mathcal{M}_2)$ represent independent subsystems of $\mathfrak{A}(\mathcal{N})$ in the sense of Sect. 5.2.2 (Einstein causality).
5. Let $\chi : \mathcal{M} \to \mathcal{N}$ be a morphism such that its image contains a Cauchy surface of $\mathcal{N}$. Then α_χ is an isomorphism (Time slice axiom).

Axiom 4 means that causally separated subsystems do not influence each other. It is equivalent to a tensor structure of the functor $\mathfrak{A}$, namely $\mathfrak{Man}$ is a tensor category by the disjoint union, with the empty set as a unit object, $\mathfrak{Alg}$ has the tensor product of algebras as a tensor structure, with the set of complex numbers as a unit object. We set $\mathfrak{A}(\emptyset) = \mathbb{C}$ and $\mathfrak{A}(\mathcal{N} \otimes \mathcal{M}) = \mathfrak{A}(\mathcal{N}) \otimes \mathfrak{A}(\mathcal{M})$. If ι_i denotes the natural embedding of spacetime $\mathcal{N}_i$ into the disjoint union $\mathcal{N}_1 \otimes \mathcal{N}_2$, then

$$\alpha_{\iota_1}(A_1) = A_1 \otimes 1 \,, \ \alpha_{\iota_2}(A_2) = 1 \otimes A_2 \,, \ A_i \in \mathfrak{A}(\mathcal{N}_i) \,, \ i = 1, 2. \tag{5.16}$$

The crucial observation is now that a causality-preserving embedding χ of a disjoint union $\mathcal{N}_1 \otimes \mathcal{N}_2$ maps the components $\mathcal{N}_1, \mathcal{N}_2$ into causally disjoint subregions $\chi \circ \iota_1(\mathcal{N}_1), \chi \circ \iota_2(\mathcal{N}_2)$. Hence we obtain the following theorem

Theorem 1. *Let $\mathfrak{A}$ be a tensor functor, i.e., for morphisms $\chi_i : \mathcal{N}_i \to \mathcal{M}_i$, $i = 1, 2$ we have*

$$\alpha_{\chi_1 \otimes \chi_2} = \alpha_{\chi_1} \otimes \alpha_{\chi_2}. \tag{5.17}$$

Then $\mathfrak{A}$ satisfies Einstein causality. On the other hand, let $\mathfrak{A}$ be defined only on connected spacetimes and assume that it satisfies Einstein causality. Then it can be uniquely extended to a tensor functor on spacetimes with finitely many connected components.

Axiom 5 may be understood as a consequence of the existence of a dynamical law which has the features of a hyperbolic differential equation with a well-posed Cauchy problem. It relates to cobordisms of Lorentzian manifolds. Namely, we may associate with a Cauchy surface $\Sigma \subset \mathcal{M}$ the inverse limit of algebras $\mathfrak{A}(\mathcal{N})$, $\Sigma \subset \mathcal{N} \subset \mathcal{M}$. The inverse limit is constructed in the following way. We consider families $(A_\mathcal{N})$, indexed by spacetimes $\mathcal{N}$ with $\Sigma \subset \mathcal{N} \subset \mathcal{M}$, which satisfy the condition

$$\alpha_{\mathcal{N}_1 \mathcal{N}_2}(A_{\mathcal{N}_2}) = A_{\mathcal{N}_1}. \tag{5.18}$$

where $\mathcal{N}_1 \mathcal{N}_2$ denotes the embedding $\mathcal{N}_2 \subset \mathcal{N}_1$. Two such families are called equivalent if they coincide for sufficiently small spacetimes. The algebra $\mathfrak{A}(\Sigma)$ is now defined as the algebra generated by these equivalence classes. By $\alpha_{\mathcal{M}\Sigma}(A) = \alpha_{\mathcal{M}\mathcal{N}}(A_\mathcal{N})$ one defines a homomorphism from $\mathfrak{A}(\Sigma)$ into $\mathfrak{A}(\mathfrak{M})$. The construction described above can be done for every submanifold. We now use the time slice axiom. Due to this axiom, the homomorphisms $\alpha_{\mathcal{N}_1 \mathcal{N}_2}$ are invertible. As a consequence, $\alpha_{\mathcal{M}\Sigma}$ is an isomorphism. Therefore, one obtains a time evolution as a propagation between Cauchy surfaces, namely the propagation from Σ to another Cauchy surface Σ' is described by the isomorphism

$$\alpha_{\Sigma'\Sigma} = \alpha_{\mathcal{M}\Sigma'}^{-1} \alpha_{\mathcal{M}\Sigma}. \tag{5.19}$$

This solves a longstanding problem dating back to ideas of Schwinger who postulated a generally covariant form of the Schrödinger equation. In its original form, as a unitary map between Hilbert spaces it cannot be realized even for free fields on Minkowski space. But in the sense of algebraic isomorphisms it always holds, provided the time slice axiom is satisfied.

In perturbation theory one wants to change the dynamical law. During the construction of the theory it turns out to be fruitful to relax the conditions so that the time slice axiom does not hold (off-shell formalism). The new dynamical law then defines an ideal of the algebra, such that the quotient again satisfies the time slice axiom.

In algebraic quantum field theory, one considers a net of subalgebras labeled by subregions of a given spacetime and requires that the net satisfies certain axioms, the Haag–Kastler, axioms. The formalism above is a proper generalization in the following sense. If we restrict our functor to the globally hyperbolic subregions

$\mathcal{N}$ of a given globally hyperbolic spacetime $\mathcal{M}$, we obtain a net of subalgebras $(\alpha_{\mathcal{MN}}(\mathfrak{A}(\mathcal{N}))$ with the proper inclusions. Moreover, isometries of $\mathcal{M}$ immediately induce further embeddings, such that the functoriality of $\mathfrak{A}$ implies covariance under symmetries. Clearly Axioms 4 and 5 correspond to the locality and the primitive causality axioms of the Haag–Kastler framework.

5.3.2 Fields as Natural Transformations

In quantum field theory fields are defined as distributions with values in the algebra of observables. They are required to transform covariantly under isometries of spacetime. On a first sight, it seems that the latter requirement becomes empty on generic spacetimes. Moreover, it seems to be difficult to compare fields which are defined on different spacetimes. But it turns out that the locally covariant framework offers the possibility for a new concept of fields. The idea is that fields have to be defined simultaneously on all spacetimes in a coherent way, namely fields may be defined as natural transformations between a functor, say $\mathcal{D}$, that associates with each spacetime $\mathcal{M}$ a space of test functions $\mathcal{D}(\mathcal{M})$ and the previous functor of a specific locally covariant theory. Here $\mathcal{D}\chi$ for an embedding $\chi : \mathcal{N} \to \mathcal{M}$ is the pushforward χ_* which is defined on functions with compact support by

$$\chi_* f(x) = \begin{cases} f(\chi^{-1}(x)) \,, \; x \in \chi(\mathcal{N}) \\ 0 \quad , \quad \text{else} \end{cases}, \tag{5.20}$$

thus $\mathcal{D}$ is a covariant functor. A natural transformation Φ from $\mathcal{D}$ to $\mathfrak{A}$ is a family $(\Phi_M)_{M \in \mathfrak{Man}}$ of linear maps $\Phi_M : \mathcal{D}(\mathcal{M}) \to \mathfrak{A}(\mathcal{M})$ which satisfy the following commutative diagram:

$$
\begin{array}{ccc}
\mathcal{D}(\mathcal{M}) & \xrightarrow{\Phi_M} & \mathfrak{A}(\mathcal{M}) \\
\chi_* \downarrow & & \downarrow \alpha_\chi \\
\mathcal{D}(\mathcal{N}) & \xrightarrow{\Phi_N} & \mathfrak{A}(\mathcal{N})
\end{array}
.
$$

The commutativity of the diagram means that the field $\Phi \equiv (\Phi_M)_{M \in \mathfrak{Man}}$ has the covariance property

$$\alpha_\chi \circ \Phi_M = \Phi_N \circ \chi_*.$$

In case χ is an isometry of a given spacetime, this reduces to the standard covariance condition for quantum fields.

The covariance condition immediately implies that the field, restricted to a small neighborhood of a given point, can depend only on the metric within the same neighborhood. Together with some more technical conditions, this was used in [7] to prove that fields can be uniquely fixed on all spacetimes by finitely many parameters. This allows a comparison of states on different spacetimes in terms of expectation values of locally covariant fields.

Also other structures in quantum field theory can be understood in terms of natural transformations. In particular one can relax the linearity condition. The naturality requirement turns out to be the crucial condition which restricts the ambiguity of the renormalization procedure.

5.4 Classical Field Theory

Before entering the somewhat involved problems of quantum field theory, we want to demonstrate that many of the general structures are already present in classical field theory. As discussed in Sect. 5.2, this amounts to the replacement of associative complex algebras by real Poisson algebras.

5.4.1 Classical Observables

Let φ be a scalar field on a globally hyperbolic spacetime $\mathcal{M}$. The space of smooth field configurations is denoted by $\mathfrak{C}(\mathcal{M}) := C^\infty(\mathcal{M})$. $\mathfrak{C}$ may be considered as a contravariant functor by identifying $\mathfrak{C}\chi$ for an embedding χ with the pullback $\chi^* h = h \circ \chi$. The basic observables are the evaluation functionals

$$\varphi(x)(h) = h(x), \quad h \in \mathfrak{C}(\mathcal{M}). \tag{5.21}$$

More generally, we consider spaces of maps $F : \mathfrak{C}(\mathcal{M}) \to \mathbb{C}$ which transform covariantly under embeddings, $\chi_* F(\varphi) = F(\varphi \circ \chi)$. We associate with each map $F : \mathfrak{C}(\mathcal{M}) \to \mathbb{C}$ a closed set $\mathrm{supp}(F)$ in analogy to the convention for distributions by

$$\mathrm{supp}(F) = \{x \in \mathcal{M} | \forall \text{neighborhoods } U \text{ of } x \ \exists \varphi, h \in \mathfrak{C}(\mathcal{M}), \mathrm{supp} h \subset U$$
$$\text{such that} F(\varphi + h) \neq F(\varphi)\}. \tag{5.22}$$

We require that these maps have compact support and are differentiable in the sense that for every $\varphi, h \in \mathfrak{C}(\mathcal{M})$ the function $\lambda \mapsto F(\varphi + \lambda h)$ is infinitely often differentiable and the nth derivative at $\lambda = 0$ is for every φ a symmetric distribution $F^{(n)}(\varphi)$ on $\mathcal{M}^n$, such that

$$\frac{d^n}{d\lambda^n} F(\varphi + \lambda h)|_{\lambda=0} = \langle F^{(n)}(\varphi), h^{\otimes n} \rangle. \tag{5.23}$$

Note that these distributions automatically have compact support with

$$\mathrm{supp} F^{(n)}(\varphi) \subset (\mathrm{supp} F)^n. \tag{5.24}$$

Moreover, $F^{(n)}$, as a map on $\mathfrak{C}(\mathcal{M}) \times \mathcal{C}^\infty(\mathcal{M}^n)$, is required to be continuous (see [8] for an introduction to these mathematical notions).

In addition we have to impose conditions on the wave front sets of the functional derivatives (see page 98 for the definition of wave front sets). Here we use different options:

$$\mathcal{F}_0(\mathcal{M}) = \{F \text{ differentiable with compact support}, \text{WF}(F^{(n)}(\varphi)) = \emptyset\}. \quad (5.25)$$

An example for such an observable is

$$F(\varphi) = \frac{1}{n!} \int d\mathrm{vol}_n \, f(x_1, \ldots, x_n)\varphi(x_1) \cdots \varphi(x_n),$$

with a symmetric test function $f \in \mathcal{D}(\mathcal{M}^n)$, with the functional derivatives

$$\langle F^{(k)}(\varphi), h^{\otimes k} \rangle = \frac{1}{k!} \int d\mathrm{vol}_n \, f(x_1, \ldots, x_n)h(x_1) \cdots h(x_k)\varphi(x_{k+1}) \cdots \varphi(x_n).$$
$$(5.26)$$

This class unfortunately does not contain the most interesting observables, namely the nonlinear local ones. We call a map F local, if it satisfies the following additivity relation for $\varphi, \psi, \chi \in \mathfrak{C}(\mathcal{M})$ with $\mathrm{supp}\varphi \cap \mathrm{supp}\chi = \emptyset$:

$$F(\varphi + \psi + \chi) = F(\varphi + \psi) - F(\psi) + F(\psi + \chi)). \quad (5.27)$$

For differentiable maps F this condition immediately implies that all functional derivatives $F^{(n)}(\varphi)$ have support on the thin diagonal

$$\Delta_n := \{(x_1, \ldots, x_n) \in \mathcal{M}^n, x_1 = \cdots = x_n\}. \quad (5.28)$$

In particular the wave front sets for $n \geq 2$ cannot be empty for $F^{(n)} \neq 0$. The best we can require is that their wave front sets are orthogonal to the tangent bundle of the thin diagonal, considered as a subset of the tangent bundle of $\mathcal{M}^n$. A simple example is $F = \frac{1}{2} \int d\mathrm{vol} \, f(x)\varphi(x)^2$ with a test function $f \in \mathcal{D}(\mathcal{M})$ where the second functional derivative at the origin is

$$\langle F^{(2)}(0), h \rangle = \int d\mathrm{vol} \, f(x)h(x, x). \quad (5.29)$$

The set of local functionals which are compactly supported, infinitely differentiable, and have wave front sets orthogonal to the tangent bundle of the thin diagonal is denoted by $\mathcal{F}_{\mathrm{loc}}(\mathcal{M})$. The set of local functionals contains in particular the possible interactions.

Examples for local functionals can be given in terms of functions on the jet bundle,

$$F(\varphi) = \int d\text{vol} f(j_x(\varphi)), \tag{5.30}$$

where $j_x(\varphi) = (x, \varphi(x), \nabla\varphi(x), \dots)$. Actually, every $F \in \mathcal{F}_{\text{loc}}$ is of this form [9]

Theorem 2. *Let $F \in \mathcal{F}_{\text{loc}}$. Then there exists a function f on the jet bundle such that (5.30) holds.*

Proof. By the fundamental theorem of calculus we have

$$F(\varphi) = F(0) + \int d\lambda \langle F^{(1)}(\lambda\varphi), \varphi \rangle. \tag{5.31}$$

By the assumption on the wave front set of F, the first derivative is a test function $x \mapsto F^{(1)}(\lambda\varphi)(x)$ with compact support. We have to prove that the value of this function at any point x depends only on the jet of φ at the point x. Let h be a test function with vanishing derivatives at x. Again from the fundamental theorem of calculus we get

$$F^{(1)}(\lambda(\varphi + h))(x) - F^{(1)}(\lambda\varphi)(x) = \int d\mu \langle F^{(2)}(\lambda(\varphi + \mu h))(x), \lambda h \rangle. \tag{5.32}$$

But $F^{(2)}(\lambda(\varphi + \mu h))$ is a distribution with support on the diagonal with wave front set orthogonal to the tangent bundle of the diagonal, thus in a chart near x it is a finite derivative of a δ distribution in the difference variables with smooth coefficients. Hence the right-hand side of (5.32) vanishes.

The space $\mathcal{F}_{\text{loc}}(\mathcal{M})$ is not closed under products. We therefore have to introduce a larger set. We choose a set which will turn out to be closed not only under the classical (pointwise) product but also under the other products we want to introduce, namely the Poisson bracket and the associative product of quantum physics. Moreover it will contain the (renormalized) time-ordered products of local functionals which are needed for the perturbative construction of interactions. These products are defined in terms of functional derivatives multiplied by Green's functions of normal hyperbolic differential operators. One therefore has to choose functionals whose derivatives have wave front sets which allow the multiplication by Green's functions. A convenient condition on the wave front sets is that they contain no covector $(x_1, \dots, x_n; k_1, \dots, k_n)$ where all k_i are elements of the closed forward light cone $\overline{V}_+(x_i)$ over the base point $x_i \in \mathcal{M}$ or all of them belong to the respective past light cones. Let $\overline{V}_\pm = \{(x, k) \in T^*(\mathcal{M}) | k \in \overline{V}_\pm(x)\}$. We then set

$$\mathcal{F}(\mathcal{M}) = \{F \text{ differentiable with compact support,}$$
$$\text{WF}(F^{(n)}(\varphi)) \cap ((\overline{V}_+^n \cup \overline{V}_-^n)) = \emptyset\}. \tag{5.33}$$

This set contains in particular the local functionals. The condition on the wave front sets will turn out to be crucial in quantum field theory.

5.4.2 Classical Dynamics

The dynamics of a classical field theory is usually given in terms of an action, e.g., $S_0(\varphi) = \int d\text{vol}\, \mathcal{L} \circ j_x(\varphi)$, with

$$\mathcal{L} = \frac{1}{2} g(d\varphi, d\varphi) - V(\varphi), \tag{5.34}$$

where V is a smooth real function. But S_0 is, for noncompact spacetimes $\mathcal{M}$, not defined for all $\varphi \in \mathcal{C}(\mathcal{M})$; we therefore multiply $\mathcal{L}$ by a test function $f \in \mathcal{D}(\mathcal{M})$ which is identical to 1 in a given relatively compact open region $\mathcal{N}$ and obtain an element $\mathcal{L}(f) = \int d\text{vol}\, f(x)\mathcal{L} \circ j_x(\varphi) \in \mathcal{F}_{\text{loc}}(\mathcal{M})$. We then take the functional derivative of the modified action $\mathcal{L}(f)$, restrict it to $\mathcal{N}$, and obtain the field equation (Euler–Lagrange equation), (which is independent of the choice of f)

$$0 = \mathcal{L}^{(1)}(\varphi) = \frac{\partial \mathcal{L}}{\partial \varphi} - \nabla_\mu \frac{\partial \mathcal{L}}{\partial \nabla_\mu \varphi} = -\Box \varphi - V'(\varphi). \tag{5.35}$$

Since $\mathcal{N}$ was arbitrary the field equation has the same form everywhere within $\mathcal{M}$.

The field equation may be linearized around an arbitrary field configuration φ. This amounts to the computation of the second derivative of the action. Again we restrict ourselves to a relatively compact open subregion $\mathcal{N}$ and determine the second derivative of $\mathcal{L}(f)$, $f \equiv 1$ on $\mathcal{N}$. The second derivative then may be understood as a differential operator which in the example above takes the form

$$\mathcal{L}^{(2)}(\varphi)h(x) = (-\Box - V''(\varphi(x)))h(x). \tag{5.36}$$

We will only consider classical actions S such that the second derivative is a normal hyperbolic differential operator and thus according to Theorem 4 on page 78 possesses unique retarded and advanced Green's functions $\Delta_S^{R,A}$.

5.4.3 Classical Møller Operators

If the action is a quadratic function, the field equation is linear and may be solved in terms of the Green functions. We now want to interpolate between different actions S which differ by an element in $\mathcal{F}(\mathcal{M})$, in analogy to quantum mechanical scattering theory where isometries (the famous Møller operators) are constructed which intertwine the interacting Hamiltonian, restricted to the scattering states, with the free Hamiltonian. We interpret $S^{(1)}$ as a map from $\mathcal{C}(\mathcal{M})$ to $\mathcal{E}'(\mathcal{M})$. We want to construct maps $r_{S_1 S_2}$ (the retarded Møller operators) from $\mathcal{C}(\mathcal{M})$ to itself with the properties

$$S_1^{(1)} \circ r_{S_1 S_2} = S_2^{(1)} ; \tag{5.37}$$

$$r_{S_1 S_2}(\varphi)(x) = \varphi(x) , \quad x \notin J_+(\text{supp}(S_1 - S_2)). \tag{5.38}$$

We set $S_1 = S + \lambda F$, $S_2 = S$, and differentiate (5.37) with respect to λ. Let $\varphi_\lambda = r_{S+\lambda F,S}(\varphi)$. We obtain

$$\langle (S + \lambda F)^{(2)}(\varphi_\lambda), \frac{d}{d\lambda}\varphi_\lambda \otimes h \rangle + \langle F^{(1)}(\varphi_\lambda), h \rangle = 0. \tag{5.39}$$

Together with condition (5.38) this implies that the Møller operators satisfy the differential equation

$$\frac{d}{d\lambda}\varphi_\lambda = -\Delta^R_{S+\lambda F}(\varphi_\lambda)F^{(1)}(\varphi_\lambda). \tag{5.40}$$

This equation has a unique solution in terms of a formal power series in λ. Moreover, by the Nash–Moser theorem, one can show that solutions exist for small λ ([9], to appear).

5.4.4 Peierls Bracket

The Møller operators can be used to endow the algebra of functionals with a Poisson bracket. This was first proposed by Peierls [10], a complete proof was given much later by Marolf [11] (see also [12]). One first defines the retarded product of two functionals F and G by

$$R_S(F, G) = \frac{d}{d\lambda} G \circ r_{S+\lambda F,G}|_{\lambda=0}. \tag{5.41}$$

From (5.40) we obtain the explicit formula in terms of the retarded Green function

$$R_S(F, G) = -\langle G^{(1)}, \Delta^R_S F^{(1)} \rangle. \tag{5.42}$$

The Peierls bracket is then a measure for the mutual influence of two possible interactions

$$\{F, G\}_S = R_S(F, G) - R_S(G, F) = \langle F^{(1)}, \Delta_S G^{(1)} \rangle, \tag{5.43}$$

with the commutator function $\Delta_S = \Delta^R_S - \Delta^A_S$.

In Peierls original formulation the functionals were restricted to solutions of the Euler–Lagrange equations for S. It is then difficult to prove the Jacobi identity. Peierls does not give a general proof and shows instead that his bracket coincides in typical cases with the Poisson bracket in a Hamiltonian formulation.

In the off-shell formalism presented above the proof of the Jacobi identity is straightforward. It relies on the formula for the functional derivative of the retarded propagator

$$\langle\langle f, \Delta^R_S g \rangle^{(1)}, h \rangle = -\langle S^{(3)}, \Delta^A_S f \otimes \Delta^R_S g \otimes h \rangle, \tag{5.44}$$

which holds since the retarded propagator is an inverse of the operator associated with $S^{(2)}$, the corresponding formula for the advanced propagator and the symmetry of the third derivative of S as a trilinear functional.

In our off-shell formalism the Peierls bracket has the form

$$\{F, G\}_S = \langle F^{(1)}, \Delta_S G^{(1)} \rangle, \tag{5.45}$$

with the commutator function $\Delta_S = \Delta_S^R - \Delta_S^A$.

The triple $(\mathcal{F}(\mathcal{M}), S, \{\cdot, \cdot\}_S)$ is termed Poisson algebra over S.

Let now $J_S(\mathcal{M})$ be the ideal (with respect to the pointwise product) in $\mathcal{F}(\mathcal{M})$ which vanishes on solutions of the field equation,

$$J_S(\mathcal{M}) = \{F \in \mathcal{F}(\mathcal{M}) | F(\varphi) = 0 \text{ whenever } S^{(1)}(\varphi) \equiv 0\}. \tag{5.46}$$

We want to prove that $J_S(\mathcal{M})$ is also an ideal for the Poisson bracket.

Theorem 3. *Let $F \in J_S(\mathcal{M})$ and $G \in \mathcal{F}(\mathcal{M})$. Then $\{F, G\}_S \in J_S(\mathcal{M})$.*

Proof. Let $\varphi \in \mathfrak{C}(\mathcal{M})$ be a solution of the field equation, i.e., $S^{(1)}(\varphi)(x) = 0 \ \forall x \in \mathcal{M}$. We want to construct a one-parameter family of solutions $\varphi_t \in \varphi_t \in \mathfrak{C}(\mathcal{M})$, $t \in \mathbb{R}$ which satisfy the initial condition $\varphi_0 = \varphi$, and the differential equation

$$\frac{d}{dt}\varphi_t = \Delta_S(\varphi_t)G^{(1)}(\varphi). \tag{5.47}$$

Provided such a solution exists, φ_t is a solution of the field equation since $S^{(1)}(\varphi_0) = S^{(1)}(\varphi) = 0$ and

$$\frac{d}{dt}S^{(1)}(\varphi_t) = S^{(2)}(\varphi_t)\frac{d}{dt}\varphi_t \tag{5.48}$$

as $\frac{d}{dt}\varphi_t$ is by construction a solution of the linearized field equation at φ_t. Then $F(\varphi_t) = 0 \ \forall t$ and $0 = \frac{d}{dt}F(\varphi_t)|_{t=0} = \{F, G\}_S(\varphi)$. It remains to show that the differential equation (5.47) has a solution. This follows in the same way as the proof of existence of local solutions for the Møller operators in (5.40).

The theorem allows to define the on-shell Poisson algebra by

$$\mathcal{F}_S(\mathcal{M}) = \mathcal{F}(\mathcal{M})/J_S(\mathcal{M}). \tag{5.49}$$

5.4.5 Local Covariance for Classical Field Theory

We want to show that classical field theory is locally covariant provided the action S is induced by a locally covariant field.

Let $\mathcal{F}$ denote the functor which associates with every $\mathcal{M} \in \mathfrak{Man}$ the commutative algebra of functionals $\mathcal{F}(\mathcal{M})$ defined before and to every morphism $\chi : \mathcal{M} \to \mathcal{N}$ the transformation

$$\mathcal{F}\chi(F)(\varphi) = F(\varphi \circ \chi). \tag{5.50}$$

Since χ preserves the metric and the time orientation, forward and backward light cones in the cotangent bundles transform properly. Together with the covariance of the wave front sets this implies that $\mathcal{F}\chi$ maps $\mathcal{F}(\mathcal{M})$ into $\mathcal{F}(\mathcal{N})$.

Let now $\mathcal{L}$ be a natural transformation from $\mathcal{D}$ to $\mathcal{F}$, i.e., for every $\mathcal{M} \in \mathfrak{Man}$ we have a linear map $\mathcal{L}_{\mathcal{M}} : \mathcal{D}(\mathcal{M}) \to \mathcal{F}(\mathcal{M})$ which satisfies

$$\mathcal{L}_{\mathcal{M}}(f)(\varphi \circ \chi) = \mathcal{L}_{\mathcal{N}}(\chi_* f)(\varphi). \tag{5.51}$$

Typical examples are given in terms of smooth functions L of two real variables by $\mathcal{L}_{\mathcal{M}}(f)(\varphi) = \int d\text{vol}_{\mathcal{M}} f(x) L(\varphi(x), g_{\mathcal{M}}(d\varphi(x), d\varphi(x)))$.

Theorem 4. $\mathcal{L}_{\mathcal{M}}(f)$ *is local, i.e., satisfies the additivity condition (5.27).*

Proof. We first show that $\text{supp}\mathcal{L}_{\mathcal{N}}(f) \subset \text{supp} f$. Let $\text{supp} h \cap \text{supp} f = \emptyset$ and let $\text{supp} f \subset \mathcal{N}$ with $\mathcal{N} \cap \text{supp} h = \emptyset$. Then from (5.51) we have

$$\mathcal{L}_{\mathcal{M}}(f)(\varphi + h) = \mathcal{L}_{\mathcal{N}}(f)((\varphi + h)|_{\mathcal{N}}) = \mathcal{L}_{\mathcal{N}}(f)(\varphi|_{\mathcal{N}}), \tag{5.52}$$

which proves the claim on the support of $\mathcal{L}_{\mathcal{M}}(f)$. Let now $\varphi, \psi, \chi \in \mathfrak{C}(\mathcal{M})$ with $\text{supp}\varphi \cap \text{supp}\chi = \emptyset$. Due to linearity in f we may decompose $\mathcal{L}_{\mathcal{M}}(f)$ into a sum of terms, each of which has disjoint support either with φ or with χ. In both cases the additivity is an immediate consequence of the support properties.

The functional derivatives of $\mathcal{L}_{\mathcal{M}}$ are defined as distributions on $\mathcal{M}^n$ which coincide on $\mathcal{N}^n$ for relatively compact open subregions $\mathcal{N}$ with the functional derivatives of $\mathcal{L}_{\mathcal{M}}(f)$ for test functions f which are identical to 1 on the subregion $\mathcal{N}$. As before, the first derivative defines the field equation $\mathcal{L}^{(1)}$ and the second functional derivative is supposed to be a normal hyperbolic differential operator. We then can equip $\mathcal{F}(\mathcal{M})$ with the Peierls bracket (5.45) and obtain a functor $\mathcal{F}_{\mathcal{L}}$ from $\mathfrak{Man}$ to the category $\mathfrak{Poi}$ of Poisson algebras which satisfies Axioms 1–4 of locally covariant quantum field theory, where the Poisson bracket on a tensor product is defined by

$$\{F_1 \otimes F_2, G_1 \otimes G_2\} = \{F_1, G_1\} \otimes F_2 G_2 + F_1 G_1 \otimes \{F_2, G_2\}. \tag{5.53}$$

The field equation defines Poisson ideals $J_{\mathcal{L}}(\mathcal{M}) \subset \mathcal{F}_{\mathcal{L}}(\mathcal{M})$ which transform under embeddings as

$$\mathcal{F}_{\mathcal{L}}\chi J_{\mathcal{L}}(\mathcal{M}) \subset J_{\mathcal{L}}(\mathcal{N}), \tag{5.54}$$

since solutions φ on $\mathcal{N}$ always induce solutions $\varphi \circ \chi$ on $\mathcal{M}$. For nonlinear field equation there may be, however, also solutions on $\mathcal{M}$ which are not of this form.

We thus obtain another functor $(\mathcal{F}/J)_{\mathcal{L}}$ which describes the on-shell theory where the field equation is satisfied. In this functor, however, the morphisms of the category of Poisson algebras are homomorphisms which are not, in general, injective.

It would be interesting to check whether this theory satisfies the time slice axiom where, in view of the possible noninjectivity of homomorphisms, isomorphy is replaced by surjectivity.

5.5 Quantum Field Theory

5.5.1 Interpretation of Locally Covariant QFT

We now turn to quantum field theory. A model of quantum field theory here is understood as a functor from the category $\mathfrak{Man}$ of globally hyperbolic spacetimes to the category of unital *-algebras which satisfies the axioms of Sect. 5.3. Our formalism differs from the formalism which may be found in standard text books for quantum field theory in Minkowski space which is either based on a representation of fields by operator-valued distribution on Fock space (canonical formulation) or on the path integral. These formulations suffer from several unsolved mathematical problems; the main reason, however, for our preference of the algebraic formulation of quantum field theory is that the concepts on which the standard approach is based lose their distinguished meaning on generic globally hyperbolic spacetimes. This can be made mathematically precise in the language of category theory by the absence of corresponding natural transformations. On a first sight, the path integral seems to be better behaved since its naive formulation involves only the classical action and the Lebesgue integral over the configuration space. The nonexistence of the Lebesgue integral on infinite-dimensional vector spaces, however, requires a choice of the Feynman propagator which is in conflict with the principle of local covariance.

On Minkowski space, the standard interpretation of the theory is based on the notion of a vacuum state and of associated excitations which are interpreted as particle states. Once a ground state is known, the interpretation of the theory in terms of cross sections is completely fixed. This was shown long ago by Araki and Haag [2] and is the basis for modern approaches to the infrared problem [13, 14]. A crucial ingredient in this analysis is the possibility to compare observables at different positions by the use of translation symmetry.

One of the main concerns for the interpretation of the theory on curved spacetimes is the absence of natural states. Here a natural state is defined as a family of states ω_M on $\mathfrak{A}(M)$, $M \in \mathfrak{Man}$ such that

$$\omega_N \circ \alpha_\chi = \omega_M , \quad \chi : M \to N. \tag{5.55}$$

A natural state could be understood as an appropriate generalization of the concept of a vacuum state. But one can show that such a state does not exist in typical cases.

This marks the most dramatic point of departure from the traditional framework of quantum field theory.

The best one can do is to associate with each spacetime $\mathcal{M}$ a natural folium of states $S_0(\mathcal{M}) \subset S(\mathfrak{A}(\mathcal{M}))$. A folium of states on a unital *-algebra is a convex set of states which is closed under the operations $\omega \mapsto \omega_A$, $\omega_A(B) = \omega(A^*BA)/\omega(A^*A)$ for elements A, B of the algebra with $\omega(A^*A) \neq 0$. A natural folium of states is a contravariant functor S_0 such that

$$S_0\chi(\omega) = \omega \circ \alpha_\chi \,, \quad \chi : \mathcal{M} \to \mathcal{N} \,, \quad \omega \in S_0(\mathcal{N}). \tag{5.56}$$

This structure allows to endow our algebras with a suitable topology, but it does not suffice for an interpretation, since it does not allow to select single states within one folium. But there is another structure which makes possible an interpretation of the theory. These are the locally covariant fields, introduced before as natural transformations. By definition they are defined on all spacetimes simultaneously, in a coherent way. Hence states on different spacetimes can be compared in terms of their values on locally covariant fields. This can be used, for instance, for a thermal interpretation of states on spacetimes without a timelike Killing vector [15].

5.5.2 Free Scalar Field

The classical free scalar field satisfies the Klein–Gordon equation

$$(\Box + m^2 + \xi R)\varphi = 0, \tag{5.57}$$

which is the Euler–Lagrange equation for the Lagrangian

$$\mathcal{L} = \frac{1}{2}(g(d\varphi, d\varphi) - (m^2 + \xi R)\varphi^2). \tag{5.58}$$

Here R is the Ricci scalar and $m^2, \xi \in \mathbb{R}$. The Klein–Gordon operator $K = \Box + m^2 + \xi R$ possesses unique retarded and advanced propagators $\Delta^{R,A}$, since we are on globally hyperbolic spacetimes (see Theorem 4 on page 78).

The corresponding functor defining the quantum theory is constructed in the following way. For each $\mathcal{M}$ we consider the *-algebra generated by a family of elements $W_{\mathcal{M}}(f)$, $f \in \mathcal{D}_{\mathbb{R}}(\mathcal{M})$ with the relations

$$W_{\mathcal{M}}(f)^* = W_{\mathcal{M}}(-f), \tag{5.59}$$

$$W_{\mathcal{M}}(f)W_{\mathcal{M}}(g) = e^{-\frac{i\hbar}{2}\langle f, \Delta g \rangle} W_{\mathcal{M}}(f + g), \tag{5.60}$$

$$W_{\mathcal{M}}(Kf) = W_{\mathcal{M}}(0). \tag{5.61}$$

This algebra has a unit $W_{\mathcal{M}}(0) \equiv 1$ and a unique C^*-norm, and its completion is the Weyl algebra over the symplectic space $\mathcal{D}(\mathcal{M})/\mathrm{im}\,K$ with the symplectic form

$\langle f, \Delta g \rangle$. With $\alpha_\chi(W_{\mathcal{M}}(f)) = W_{\mathcal{N}}(\chi_* f)$ one obtains a functor satisfying also the Axioms 4 and 5. Moreover, $W = (W_{\mathcal{M}})$ is a (nonlinear) locally covariant field. It is, however, difficult to find other locally covariant fields for this functor.

The free field itself is thought to be related to the Weyl algebra by the formula

$$W_{\mathcal{M}}(f) = e^{i\varphi_{\mathcal{M}}(f)}. \tag{5.62}$$

This relation can be established in the so-called regular representations of the Weyl algebra, in which the one-parameter groups $W_{\mathcal{M}}(\lambda f)$ are strongly continuous. But one can also directly construct an algebra generated by the field itself. It is the unital $*$-algebra generated by the elements $\varphi_{\mathcal{M}}(f)$, $f \in \mathcal{D}(\mathcal{M})$ by the relations

$$f \mapsto \varphi_M(f) \text{ is linear,} \tag{5.63}$$

$$\varphi_M(f)^* = \varphi_M(\overline{f}), \tag{5.64}$$

$$[\varphi_M(f), \varphi_M(g)] = i\hbar \langle f, \Delta g \rangle, \tag{5.65}$$

$$\varphi_{\mathcal{M}}(Kf) = 0. \tag{5.66}$$

Again one obtains a functor which satisfies Axioms 1–5. If we omit the condition (5.66) (then the time slice axiom is no longer valid and one is on the off-shell formalism), the algebra may be identified with the space of functionals on the space of field configurations $\mathfrak{C}(\mathcal{M})$,

$$F(\varphi) = \sum_{\text{finite}} \int d\text{vol}_n \, f_n(x_1, \ldots, x_n)\varphi(x_1) \cdots \varphi(x_n), \tag{5.67}$$

where f_n is a finite sum of products of test functions in one variable and where the product is given by

$$(F \star G)(\varphi) = \sum_n \frac{i^n \hbar^n}{2^n n!} \langle F^{(n)}(\varphi), \Delta^{\otimes n} G^{(n)}(\varphi) \rangle. \tag{5.68}$$

Hence, as a vector space, it may be considered as a subspace of the space $\mathcal{F}_0(\mathcal{M})$ known from classical field theory. Moreover, the involution $A \mapsto A^*$ coincides with complex conjugation. As a formal power series in $\hbar$, the product can be extended to all of $\mathcal{F}_0(\mathcal{M})$, thus providing $\mathcal{F}_0(\mathcal{M})[[\hbar]]$ with the structure of a unital $*$-algebra.

The Poisson ideal of the classical theory which is generated by the field equation turns out to coincide with the ideal with respect to the $\star$-product.

Theorem 5. *Let* $J_0(\mathcal{M})$ *be the set of all* $F \in \mathcal{F}_0(\mathcal{M})[[\hbar]]$ *with* $F(\varphi) = 0$ *whenever* $K\varphi = 0$. *Then* $J_0(\mathcal{M})$ *is a* $\star$-*ideal.*

Proof. Let $F \in J_0(\mathcal{M})$, $G \in \mathcal{F}_0(\mathcal{M})$, and $K\varphi = 0$. By the definition of the functional derivative, the distribution $F^{(n)}(\varphi)$ vanishes on n-fold tensor products of solutions, hence on $\Delta^{\otimes n} G^{(n)}(\phi)$. Thus $F \star G \in J_0(\mathcal{M})$. This shows that $J_0(\mathcal{M})$ is a right ideal. But $J_0(\mathcal{M})$ is invariant under complex conjugation, so $(G \star F)^* = F^* \star G^*$, and it is also a left ideal.

5.5.3 The Algebra of Wick Polynomials

In order to include pointwise products of fields, or more generally, local functionals in the sense of Sect. 5.4.1 into the formalism we have to admit more singular coefficients in the expansion (5.67). But then the product may become ill-defined. As an example consider the functionals

$$F(\varphi) = \int d\text{vol}\, f(x)\varphi(x)^2, \tag{5.69}$$

$$G(\varphi) = \int d\text{vol}\, g(x)\varphi(x)^2, \tag{5.70}$$

with test functions f and g. Insertion into the formula for the product yields

$$(F * G)(\varphi)$$
$$= \int d\text{vol}_2\, f(x)g(y)\left(\varphi^2(x)\varphi^2(y) + 4i\hbar\Delta(x, y)\varphi(x)\varphi(y) - 2\hbar^2\Delta(x, y)^2\right) \tag{5.71}$$

The problematic term is the square of the distribution Δ. Here the methods of microlocal analysis enter, namely the wave front set of Δ is (Strohmaier, Theorem 16)

$$\text{WF}(\Delta) = \{(x, y; k, k'), x \text{ and } y \text{ are connected by a null geodesic} \gamma,$$
$$k\|g(\dot\gamma, \cdot), U_\gamma k + k' = 0, U_\gamma \text{ parallel transport along } \gamma\}. \tag{5.72}$$

The product of Δ cannot be defined in terms of Hörmander's criterion for the multiplication of distribution, since the sum of two vectors in the wave front set can yield zero. The crucial fact is now that Δ can be split in the form

$$\Delta = \frac{1}{2}\Delta + iH + \frac{1}{2}\Delta - iH, \tag{5.73}$$

where the "Hadamard function" H is symmetric and the wave front set of $\frac{1}{2}\Delta + iH$ contains only the positive frequency part (Strohmaier, Definition 10)

$$\text{WF}\left(\frac{1}{2}\Delta + iH\right) = \{(x, y; k, k') \in \text{WF}(\Delta), k \in \overline{V_+}\}. \tag{5.74}$$

On Minkowski space, Δ depends only on the difference $x - y$, and one may find H in terms of the Fourier transform of Δ

$$\frac{1}{2}\Delta + iH = \Delta_+, \ \tilde{\Delta}_+(k) = \begin{cases} \tilde{\Delta}(k), & k \in \overline{V_+} \\ 0, & \text{else} \end{cases}. \tag{5.75}$$

On a generic spacetime, the split (5.73) represents a microlocal version of the decomposition into positive and negative energies (microlocal spectrum condition [5]) which is fundamental for quantum field theory on Minkowski space.

If we replace in the definition of the product (5.68) Δ by $\Delta + 2iH$, we obtain a new product $\star_H$. On $\mathcal{F}_0(\mathcal{M})[[\hbar]]$ this product is equivalent to $\star$, namely

$$F \star_H G = \alpha_H(\alpha_H^{-1}(F) \star \alpha_H^{-1}(G)), \tag{5.76}$$

where

$$\alpha_H(F) = \sum \frac{\hbar^n}{n!} \langle H^{\otimes n}, F^{(2n)} \rangle \tag{5.77}$$

is a linear isomorphism of $\mathcal{F}_0(\mathcal{M})[[\hbar]]$ with inverse $\alpha_H^{-1} = \alpha_{-H}$.

This product now yields well-defined expressions in (5.71); actually, it is well defined on $\mathcal{F}(\mathcal{M})[[\hbar]]$. This is a consequence of Hörmander's criterion for the multiplicability of distributions, namely by the microlocal spectrum condition (5.74) the wave front set of $(\Delta + 2iH)^{\otimes n}$ is contained in $\overline{V}_+^n \times \overline{V}_-^n$. Hence by the condition on the wave front set of the nth derivatives of $F, G \in \mathcal{F}(\mathcal{M})$ the pointwise product of the distribution $F^{(n)} \otimes G^{(n)}$ with $(\Delta + 2iH)^{\otimes n}$ exists and is a distribution with compact support. Therefore the terms in the formal power series defining the $*$-product are well defined. Moreover, they are again elements of $\mathcal{F}(\mathcal{M})$. This follows from the fact that the derivatives of $\langle F^{(n)}, (\Delta + 2iH)^{\otimes n} G^{(n)} \rangle$ arise from contractions of the pointwise products $F^{(n+k)} \otimes G^{(n+l)}$ with $(\Delta + 2iH)^{\otimes n}$ in the joint variables.

If we restrict ourselves to polynomial functionals, i.e., those for which the functional derivatives of sufficiently high orders vanish, we may set $\hbar = 1$. Up to taking the quotient by the ideal $J_0(\mathcal{M})$ of the field equation we obtain, on Minkowski space, the algebra of Wick polynomials. We thus succeeded to define on generic spacetimes an algebra containing all local field polynomials.

The annoying feature, however, is that the product depends on the choice of H. Fortunately, the difference w between two Hadamard functions H and H' is smooth.

Theorem 6. *Let H, H' be symmetric distributions in two variables satisfying condition (5.74). Then $w = H - H'$ is smooth.*

Proof. Since $w = (H - \frac{i}{2}\Delta) - (H' - \frac{i}{2}\Delta)$, the wave front set of w satisfies also condition (5.74). Thus $(x, y; k, k') \in \text{WF}(w)$ implies $k \in \overline{V}_+(x)$. But w is symmetric, hence then also $k' \in \overline{V}_+(y)$. But $-k'$ is the parallel transport of k along a null geodesic from x to y. Since $\mathcal{M}$ is time oriented, this implies $k = k' = 0$. Since by definition, the zero covectors are not in the wave front set, the wave front set of w is empty, hence w is smooth.

The smoothness of w implies that the products $*_H$ and $*_{H'}$ are equivalent

$$F *_{H'} G = \alpha_w(\alpha_w^{-1}(F) *_H \alpha_w^{-1}(G)), \tag{5.78}$$

where α_w is defined in analogy to (5.77), but is now, due to the smoothness of w, a well-defined linear isomorphism of $\mathcal{F}(\mathcal{M})[[\hbar]]$.

In order to eliminate the influence of H we replace our functionals by families $F = (F_H)$, labeled by Hadamard functions H and satisfying the coherence condition $\alpha_w(F_H) = F_{H+w}$. The product of two such families is defined by

$$(F \star G)_H = F_H \star_H G_H. \tag{5.79}$$

We call this algebra the algebra of quantum observables and denote it by $\mathcal{A}(\mathcal{M})$. The subspace of local elements $A \in \mathcal{A}_{\mathrm{loc}}(\mathcal{M})$ is formed by families $A = (A_H)$ with $A_H \in \mathcal{F}_{\mathrm{loc}}(\mathcal{M})$. Since α_w leaves $\mathcal{F}_{\mathrm{loc}}(\mathcal{M})$ invariant, $A \in \mathcal{A}_{\mathrm{loc}}(\mathcal{M})$ if $A_H \in \mathcal{F}_{\mathrm{loc}}(\mathcal{M})$ for some Hadamard function H.

$\mathcal{F}_0(\mathcal{M})[[\hbar]]$ equipped with the product (5.68) is embedded into $\mathcal{A}(\mathcal{M})$ by

$$F \mapsto (F_H) \text{ with } F_H = \alpha_H(F). \tag{5.80}$$

One may equip $\mathcal{F}(\mathcal{M})$ with a suitable topology such that α_w is a homeomorphism and such that $\mathcal{F}_0(\mathcal{M})[[\hbar]]$ is sequentially dense in $\mathcal{A}(\mathcal{M})$ [16].

5.5.4 Interacting Models

In order to treat interactions we introduce a new product $\cdot_T$ on $\mathcal{F}_0(\mathcal{M})[[\hbar]]$, the time-ordered product. It is a commutative product which coincides with the $*$-product if the factors are time ordered:

$$F \cdot_T G = F \star G \text{ if } \mathrm{supp}(F) \gtrsim \mathrm{supp}(G), \tag{5.81}$$

where $\gtrsim$ means that there is a Cauchy surface such that the left-hand side and the right-hand side are in the future and past of the surface, respectively. For the free field, we find

$$\varphi(f) \cdot_T \varphi(g) = \varphi(f)\varphi(g) + i\hbar\langle f, \Delta^D g\rangle, \tag{5.82}$$

with the "Dirac propagator" (see [17])

$$\Delta^D = \frac{1}{2}(\Delta^R + \Delta^A). \tag{5.83}$$

The time-ordered product may be extended to all of $\mathcal{F}_0(\mathcal{M})[[\hbar]]$ by

$$(F \cdot_T G)(\varphi) = \sum_n \frac{i^n \hbar^n}{n!} \langle F^{(n)}, (\Delta^D)^{\otimes n} G^{(n)}\rangle. \tag{5.84}$$

In text books on quantum field theory, the time-ordered product is usually defined for fields in the Fock space representation. But the Dirac propagator is not a solution of the homogeneous Klein–Gordon equation. Hence $J_0(\mathcal{M})$ is not an ideal with respect to the time-ordered product. Instead from $\Delta_D K = \mathrm{id}$ one finds the relation

$$\varphi(Kf) \cdot_T F = \varphi(Kf)F + i\hbar\langle F^{(1)}, \Delta_D Kf\rangle = \varphi(Kf)F + i\hbar\langle F^{(1)}, f\rangle. \quad (5.85)$$

This relation is the prototype of the so-called Schwinger–Dyson equation by which the field equation of interacting quantum fields can be formulated in terms of expectation values of time-ordered products. Since the ideal generated by the field equation vanishes in the Fock space representation, time ordering on Fock space is not well defined as a product of operators. On $\mathcal{F}_0(\mathcal{M})[[\hbar]]$, however, it is well defined and is even equivalent to the pointwise (classical) product, namely we introduce the "time-ordering operator"

$$T F(\varphi) = \sum_n \frac{i^n \hbar^n}{n!} \langle (\Delta^D)^{\otimes n}, F^{(2n)}(\varphi)\rangle. \quad (5.86)$$

T is a linear isomorphism, with the inverse obtained by complex conjugation, and

$$F \cdot_T G = T(T^{-1}(F) \cdot T^{-1}(G)). \quad (5.87)$$

In terms of T, explicit formulae for interacting fields can be given by the use of the formal S-matrix which is just the exponential function computed via the time-ordered product

$$S(F) = T \exp(T^{-1}(F)). \quad (5.88)$$

In terms of S we can write down the analog of the Møller operators for quantum field theory, via Bogoliubov's formula

$$R_V(F) \doteq \left.\frac{d}{d\lambda} S(V)^{-1} \star S(V + \lambda F)\right|_{\lambda=0} = S(V)^{-1} \star (S(V) \cdot_T F), \quad (5.89)$$

where the inverse is built with respect to the $\star$-product. R_V is a linear map from $\mathcal{F}_0(\mathcal{M})[[\hbar]]$ to itself and describes the transition from the free action to the action with additional interaction term V. It satisfies two important conditions, retardation and equation of motion. As far as the retardation property is concerned, one observes that if $\mathrm{supp}(V) \gtrsim \mathrm{supp}(F)$, the time-ordered product and the $\star$-product coincide, hence by associativity of the $\star$-product $R_V(F) = F$, so the observable F is not influenced by an interaction which takes place in the future. We now show that the interacting field $f \mapsto R_V(\varphi(Kf))$ satisfies the off-shell field equation

$$R_V(\varphi(Kf)) = \varphi(Kf) + i\hbar R_V(\langle V^{(1)}, f\rangle), \quad (5.90)$$

where $f \in \mathcal{D}(M)$ and K is the Klein–Gordon operator. (In a more suggestive notation, the field equation above reads

$$K\varphi_V(x) = K\varphi(x) + i\left(\frac{\delta V}{\delta\varphi(x)}\right)_V, \tag{5.91}$$

with the free field φ, the interacting field φ_V, and the interacting current $i\left(\frac{\delta V}{\delta\varphi}\right)_V$.)

Proof. S is the time-ordered exponential, hence by the chain rule we obtain $\langle S(V)^{(1)}, g\rangle = S(V) \cdot_T \langle V^{(1)}, g\rangle$. From (5.85)

$$R_V(\varphi(Kf)) = S(V)^{-1} \star (S(V) \cdot_T \varphi(Kf))$$

$$= S(V)^{-1} \star \left(S(V) \cdot \varphi(Kf) + i\hbar\, S(V) \cdot_T \langle V^{(1)}, f\rangle\right).$$

But $S(V) \cdot \varphi(Kf) = S(V) \star \varphi(Kf)$ since the higher order terms in $\hbar$ of the $\star$-product vanish due to $\Delta K = 0$. The statement now follows from associativity of the $\star$-product.

5.5.5 Renormalization

The remaining problem is the extension of the time-ordered product to local functionals. Here the problem can only partially be solved by the transition to an equivalent product

$$F \cdot_{T_H} G = \alpha_H(\alpha_H^{-1}(F) \cdot_T \alpha_H^{-1}(G)). \tag{5.92}$$

This transformation amounts to replacing the Dirac propagator by the Feynman-like propagator $\Delta^D + iH$. Since $\Delta^D + iH$ coincides on the complement of the support of the advanced propagator Δ^A with $\frac{1}{2}\Delta + iH$ and on the complement of the support of the retarded propagator Δ^R with $-\frac{1}{2}\Delta + iH$, its wave front set is

$$\mathrm{WF}(\Delta^D + iH) = \{(x, y, k, k') \in \mathrm{WF}(\Delta), k \in \overline{V_\pm} \text{ if } x \in J_\pm(y)\}$$
$$\cup\{(x, x, k, -k), k \neq 0\}.$$

Thus contrary to the Dirac propagator, pointwise products of these propagators exist outside of the diagonal. The problem which remains to be solved in renormalization is therefore to extend a distribution which is defined on the complement of some submanifold (the thin diagonal in our case) to the full manifold [18].

The construction can be much simplified by the fact that the time-ordered product coincides with the product $\star$ for time-ordered supports. For local functionals the time-ordered product is therefore defined whenever the localizations are different, namely let $\mathcal{L}_i, i = 1, \ldots, n$ be Lagrangians, i.e., natural transformations in the sense of Sect. 5.4.5. Then the time-ordered product $(\mathcal{L}_1 \otimes \cdots \otimes \mathcal{L}_n)_T$ can be defined in terms of an $\mathfrak{A}(\mathcal{M})$-valued distribution on $\mathcal{M}^n \setminus D$ where D is the subset where at least two variables coincide. Indeed, on tensor products of test functions $f_1 \otimes \cdots \otimes f_n$ with $\mathrm{supp}\, f_i \gtrsim \mathrm{supp}\, f_{i+1}, i = 1, \ldots, n-1$ the time-ordered product is given by

$$(\mathcal{L}_1 \otimes \cdots \otimes \mathcal{L}_n)_T^{\mathcal{M}}(f_1 \otimes \cdots \otimes f_n) = \mathcal{L}_1^{\mathcal{M}}(f_1) \star \cdots \star \mathcal{L}_n^{\mathcal{M}}(f_n). \qquad (5.93)$$

Moreover, the time-ordered product is required to be symmetric, hence it is well defined on $\mathcal{M}^n \setminus D$.

One now proceeds by induction. The time-ordered product with one factor is the Lagrangian itself. Now assume that time-ordered products of less than n factors have been constructed in the sense of $\mathfrak{A}(\mathcal{M})$-valued distributions $(\mathcal{L}_1 \otimes \cdots \otimes \mathcal{L}_k)_T^{\mathcal{M}}$ on $\mathcal{M}^k$ such that $(\mathcal{L}_1 \otimes \cdots \otimes \mathcal{L}_k)_T$ is a natural transformation from $\mathcal{D}^{\otimes k}$ to $\mathfrak{A}$ which in particular satisfies the causality condition

$$(\mathcal{L}_1 \otimes \cdots \otimes \mathcal{L}_k)_T^{\mathcal{M}}(f \otimes g) = (\mathcal{L}_1 \otimes \cdots \otimes \mathcal{L}_l)_T^{\mathcal{M}}(f) \star (\mathcal{L}_{l+1} \otimes \cdots \otimes \mathcal{L}_k)_T^{\mathcal{M}}(g) \quad (5.94)$$

provided $\operatorname{supp}(f) \subset \mathcal{M}_1^l$, $\operatorname{supp}(g) \subset \mathcal{M}_2^{k-l}$, and $\mathcal{M}_1, \mathcal{M}_2$ are subregions of $\mathcal{M}$ with $\mathcal{M}_1 \gtrsim \mathcal{M}_2$.

We may now, on $\mathcal{M}^n \setminus \Delta_n$, use a decomposition of unity $(\chi_I)_I$, indexed by the nonempty proper subsets of $\{1, \ldots, n\}$, with supports $\operatorname{supp}\chi_I \subset U_I = \{(x_1, \ldots, x_n) \in \mathcal{M}^n | \{x_i, i \in I\} \gtrsim \{x_j, j \notin I\}\}$. Then we define

$$(\mathcal{L}_1 \otimes \cdots \otimes \mathcal{L}_n)_T^{\mathcal{M}} = \sum_I \chi_I (\mathcal{L}_1 \otimes \cdots \otimes \mathcal{L}_n)_{T,I}^{\mathcal{M}}, \qquad (5.95)$$

where $(\mathcal{L}_1 \otimes \cdots \otimes \mathcal{L}_n)_{T,I}^{\mathcal{M}}$ is determined on U_I by

$$(\mathcal{L}_1 \otimes \cdots \otimes \mathcal{L}_n)_{T,I}^{\mathcal{M}}(f_1 \otimes \cdots \otimes f_n) = (\otimes_{i \in I} \mathcal{L}_i)_T^{\mathcal{M}}(\otimes_{i \in I} f_i) * (\otimes_{j \notin I} \mathcal{L}_j)_T^{\mathcal{M}}(\otimes_{j \notin I} f_i). \qquad (5.96)$$

This definition does not depend on the choice of the decomposition of unity. This follows from the fact that on intersections $U_I \cap U_J$ the distributions $(\mathcal{L}_1 \otimes \cdots \otimes \mathcal{L}_n)_{T,I}^{\mathcal{M}}$ and $(\mathcal{L}_1 \otimes \cdots \otimes \mathcal{L}_n)_{T,J}^{\mathcal{M}}$ coincide.

The crucial step now is the extension of these distributions to the full space $\mathcal{M}^n$ such that the causality condition (5.94) is satisfied. This can be done [18], but the process is, in general, not unique.

As a result we obtain a renormalized S-matrix S as a generating functional for time-ordered products

$$S(\underline{\mathcal{L}}^{\mathcal{M}}(\underline{f})) = \sum \frac{1}{n!} (\mathcal{L}_{i_1} \otimes \cdots \otimes \mathcal{L}_{i_n})_T^{\mathcal{M}}(f_{i_1} \otimes \cdots \otimes f_{i_n}), \qquad (5.97)$$

with

$$\underline{\mathcal{L}}^{\mathcal{M}}(\underline{f}) = \sum \mathcal{L}_i(f_i). \qquad (5.98)$$

The crucial conditions that restrict the ambiguities in the extension process is now that S satisfies the causality condition

$$S(\underline{\mathcal{L}}^{\mathcal{M}}(\underline{f} + \underline{g})) = S(\underline{\mathcal{L}}^{\mathcal{M}}(\underline{f})) \star S(\underline{\mathcal{L}}^{\mathcal{M}}(\underline{g})) \qquad (5.99)$$

as a consequence of (5.94) and the naturality condition

$$\alpha_\chi S(\underline{\mathcal{L}}^{\mathcal{M}}(\underline{f})) = S(\underline{\mathcal{L}}^{\mathcal{N}}(\chi_* \underline{f})) \qquad (5.100)$$

as a consequence of the naturality conditions on the time-ordered products of Lagrangians. These conditions imply the *Main Theorem of Renormalization*:

Theorem 7. *Let S_i be two extensions of the formal S-matrix to $\mathcal{A}_{\text{loc}}$ fulfilling the causality and naturality conditions. Then there exists a uniquely determined natural equivalence $Z : \mathcal{A}_{\text{loc}}[\![\hbar]\!] \to \mathcal{A}_{\text{loc}}[\![\hbar]\!]$ (a formal diffeomorphism on the space of interactions) with $Z^{(1)} = \text{id}$ such that*

$$S_2 = S_1 \circ Z. \qquad (5.101)$$

The natural equivalences Z occurring in the theorem form a group, the renormalization group in the sense of Stückelberg and Petermann. Typically, additional conditions on S induce cocycles on the renormalization group and the cohomology classes of these cocycles are the famous anomalies of quantum field theory.

We conclude that a Lagrangian alone does not specify a quantum field theoretical model completely. One has in addition to fix a point of the orbit of the interaction under the renormalization group. This amounts to a choice of suitable renormalization conditions. An important class of interactions are the *renormalizable* interactions. They have the property that the orbit under the renormalization group (after imposing suitable conditions) is finite dimensional, such that the theory can be fixed in terms of finitely many parameters.

The method of renormalization described above is termed causal perturbation theory and was first rigorously performed by Epstein and Glaser on Minkowski space [19], based on previous work of Stückelberg and Bogoliubov. Its extension to curved spacetimes was undertaken by Brunetti and Fredenhagen [18], and the implementation of the principle of local covariance and the reduction to finitely many free parameters is due to Hollands and Wald [7, 20]. The extension of the method to gauge theories was performed on Minkowski space by Dütsch, Scharf et al. [21] and generalized to curved spacetimes by Hollands [22].

On Minkowski space, there exist other methods of renormalization, which are known to be equivalent. One of these is the Bogoliubov–Parasiuk–Hepp–Zimmermann method whose involved structure was recently made transparent in terms of the Connes–Kreimer Hopf algebra [23]. Another one is the Wilson–Polchinski method of renormalization group flow equations [16], where the time-ordered product is regularized. The dependence of the theory under a variation of the regularization delivers the so-called flow equation. In the sense of formal power series, the flow equation can always be solved, and the removal of the regularization amounts to asymptotic stability properties of the solutions. The attractive feature of this method is that the concepts do not depend on the perturbative formulation. It is usually defined in terms of the path integral which seems to make a formulation on curved spacetime difficult. But if interpreted not as an integral but as an integral operator, it can actually be identified with the formal S-matrix of causal perturbation theory.

Namely let T_Λ be a regularized version of the time-ordering operator T obtained by replacing the Feynman propagator $\Delta_D + iH$ by a sufficiently regular distribution $G_\Lambda + iH$. Then $S_\Lambda = T_\Lambda \circ \exp \circ T_\Lambda^{-1}$ is a well-defined generating functional for regularized time-ordered products on $\mathcal{A}$. Different regularizations may be compared in terms of the *effective action* $S_{\Lambda_1}^{-1} \circ S_{\Lambda_2}$ which yields after application to $V \in \mathcal{A}(\mathcal{M})$ a modified interaction $V_{\Lambda_1 \Lambda_2}$ which is interpreted as the "interaction at scale Λ_1 after integrating out the degrees of freedom beyond Λ_2." This interpretation refers to a regularization by a momentum cutoff and has no immediate generalization to the generic situation on curved space time. But in any case we know from causal perturbation theory [16] that given S there exist renormalization group transformations Z_Λ such that

$$S = \lim S_\Lambda \circ Z_\Lambda, \tag{5.102}$$

if $G_\Lambda + iH$ converges to $\Delta_D + iH$ in the appropriate sense (Hörmander's topology for distributions with prescribed wave front set). The renormalization transformation Z_Λ is the operation which adds the necessary counter terms to the interaction. If Λ can be identified with a complex variable such that S_Λ is meromorphic and $\Lambda = 0$ corresponds to the removal of the regularization, one can choose Z_Λ such that it removes the pole at $\Lambda = 0$ and obtains a distinguished choice for S. For instance, in the case of dimensional regularization this defines the so-called minimal renormalization. But such a choice of S is not necessarily appropriate from the point of view of physics. In particular it depends on the choice of the regularization. It can, however, be used to fix a specific point on the orbit of interactions under the renormalization group and thus allow an explicit formulation of renormalization conditions.

References

1. Haag, R., Kastler, D.: An algebraic approach to quantum field theory. J. Math. Phys. **5**, 848 (1964)
2. Haag, R.: Local Quantum Physics. 2nd edn. Springer-Verlag, Berlin, Heidelberg, New York (1996)
3. Peskin, M.E., Schröder, D.V.: An Introduction to Quantum Field Theory. Perseus (1995)
4. Brunetti, R., Fredenhagen, K., Verch, R.: The generally covariant locality principle – A new paradigm for local quantum physics, Commun. Math. Phys. **237**, 31 (2003)
5. Radzikowski, M.: Micro-local approach to the Hadamard condition in quantum field theory in curved spacetime. Commun. Math. Phys. **179**, 529 (1996)
6. Brunetti, R., Fredenhagen, K., Köhler, M.: The microlocal spectrum condition and Wick polynomials of free fields on curved spacetimes. Commun. Math. Phys. **180**, 633–652 (1996)
7. Hollands, S., Wald, R.M.: Local wick polynomials and time-ordered-products of quantum fields in curved spacetime. Commun. Math. Phys. **223**, 289 (2001)

8. Hamilton, R.S.: The inverse function theorem of Nash and Moser. Bull. (New Series) Am. Math. Soc. **7**, 65 (1982)
9. Brunetti, R., Fredenhagen, K., Ribeiro, P.L.: work in progress.
10. Peierls, R.: The commutation laws of relativistic field theory. Proc. Roy. Soc. (London) A **214**, 143 (1952)
11. Marolf, D.M.: The generalized Peierls brackets. Ann. Phys. **236**, 392 (1994)
12. deWitt, B.S.: The spacetime approach to quantum field theory. In: B.S. deWitt, R. Stora, (eds.), Relativity, Groups and Topology II: Les Houches 1983, part 2, 381, North-Holland, New York (1984)
13. Buchholz, D., Porrmann, M., Stein, U.: Dirac versus Wigner: Towards a universal particle concept in local quantum field theory. Phys. Lett. B **267**, 377 (1991)
14. Steinmann, O.: Perturbative Quantum Electrodynamics and Axiomatic Field Theory. Springer, Berlin, Heidelberg, New York (2000)
15. Buchholz, D., Ojima, I., Roos, H.: Thermodynamic properties of non-equilibrium states in quantum field theory. Ann. Phys. NY. **297**, 219 (2002)
16. Brunetti, R., Dütsch, M., Fredenhagen, K.: Perturbative Algebraic Quantum Field Theory and the Renormalization Groups. Preprint http://arxiv.org/abs/0901.2038
17. Dirac, P.A.M.: Classical theory of radiating electrons. Proc. Roy. Soc. London A**929**, 148 (1938)
18. Brunetti, R., Fredenhagen, K.: Microlocal analysis and interacting quantum field theories: Renormalization on physical backgrounds. Commun. Math. Phys. **208**, 623 (2000)
19. Epstein, H., Glaser, V.: The role of locality in perturbation theory. Annales Poincare Phys. Theor. A **19**, 211 (1973)
20. Hollands, S., Wald, R.M.: Existence of local covariant time-ordered-products of quantum fields in curved spacetime. Commun. Math. Phys. **231**, 309 (2002)
21. Duetsch, M., Hurth, T., Krahe, F., Scharf, G.: Causal Construction of Yang-Mills Theories 1. Nuovo Cim. A **106**, 1029 (1993)
22. Hollands, S.: Renormalized quantum yang-mills fields in curved spacetime. Rev. Math. Phys. **20**, 1033 (2008)
23. Connes, A., Kreimer, D.: Renormalization in quantum field theory and the Riemann-Hilbert problem. I: The Hopf algebra structure of graphs and the main theorem. Commun. Math. Phys. **210**, 249 (2000)

Index

$A^\times$, invertible elements in A, 3
$C_0(X)$, continuous functions vanishing at infinity, 2
$C_0^\infty(X)$, smooth functions vanishing at infinity, 2
$D(S)$, Cauchy development of subset S, 56
G_F, Feynman propagator, 117
$G_\pm$, Green's distributions, 122
$I_+^M(A)$, chronological future of subset A in M, 45
$I_+^M(x)$, chronological future of point x in M, 45
$I_-^M(A)$, chronological past of A, 45
$I_-^M(x)$, chronological past of x, 45
$J^M(A) = J_+^M(A) \cup J_-^M(A)$, 45
$J_+^M(A)$, causal future of subset A in M, 45
$J_+^M(x)$, causal future of point x in M, 45
$J_-^M(A)$, causal past of A, 45
$J_-^M(x)$, causal past of x, 45
$J_\pm(x)$, caucal future/past of x, 122
$R_\pm$, Riesz distributions, 114
S', commutant of S, 21
$S(A)$, states on A, 17
$W(\phi)$, Weyl-system, 29
$[\cdot, \cdot]$, commutator, 21
$\|\cdot_\iota\|$, injective C^*-norm, 24
$\|\cdot\|_\pi$, projective C^*-norm, 25
Δ_n, thin diagonal, 138
$\exp_p$, exponential map, 47
$\langle W(V)\rangle$, linear span of the $W(\phi)$, $\phi \in V$, 32
$\mathcal{D}'$, set of distributions, 87
$\mathcal{E}$, set of compactly supported smooth functions, 86
$\mathcal{E}$, set of smooth functions, 86
$\mathcal{E}'$, set of compactly supported distributions, 87
$\mathcal{S}$, set of Schwartz functions, 93
$\mathcal{S}'$, set of Schwartz distributions, 93

WF(ϕ), wavefront set of ϕ, 98
char(P), characteristic set of P, 119
CCR(S), $*$–morphism induced by a symplectic linear map, 36
CCR(V, ω), CCR-algebra of a symplectic vector space, 30, 35
$\mathcal{L}(H)$, bounded operators on Hilbert space H, 1
CCR, functor SymplVec $\to$ C^*Alg, 36
C^*Alg, category of C^*-algebras and injective $*$–morphisms, 36
$\rho_A(a)$, spectral radius of $a \in A$, 4
$\sigma_A(a)$, spectrum of $a \in A$, 4
$\Box$, wave operator, 113
$\Sigma(\phi)$, set of singular directions of a distribution, 99
$p < q$, there is a future-directed causal curve from p to q, 45
$p \leq q$, either $p < q$ or $p = q$, 45
$p \ll q$, there is a future-directed timelike curve from p to q, 45
$r_A(a)$, resolvent set of $a \in A$, 4
SymplVec, category of symplectic vector spaces and symplectic linear maps, 36
$*$-automorphism, 12
$*$-morphism, 12

A
acausal subset, 55
achronal subset, 55
advanced fundamental solution, 64
advanced Green's distribution, 122
advanced Green's operator, 81
advanced Riesz distribution, 67
anti-deSitter spacetime, 43

B
Bell inequality, 26
Bogoliubov transformation, 36

157